3600 celestial asterisms for amateur astronomers

by

Martin P Nicholson

WHAT IS AN ASTRONOMICAL ASTERISM?

Rather confusingly the word asterism has a number of different meanings to an astronomer depending on the context in which the word is being used. An asterism can be a large, sometimes very large, scale feature in the night sky. The stars making up an asterism can be all from the same constellation – for example the "belt of Orion" – or they can be from different constellations as is the case with the "summer triangle".

Increasingly the word is used to describe a group of stars that appear to be close together in the sky but where the stars have no proven link. An asterism can appear similar to an open cluster but unlike the stars in an open cluster the stars in an asterism are probably at significantly different distances from the observer.

An amusing definition of an asterism is "a group of stars that has not been categorised as something else"!

Some well-known and high profile groups such as the Astronomical League have created observing programs specifically intended to encourage amateur astronomers to observe asterisms and some amateur astronomers have invested large amounts of energy and enthusiasm in looking for asterisms to add to the ever-growing catalogue of these objects.

It is disappointing that some researchers have perhaps taken themselves and their desire to find new objects rather too seriously. Amateur astronomy is supposed to be fun and those people seeking to impose a rigid framework on what constitutes an asterism worthy of "catalogue status" need to remind themselves that any catalogue they create is entirely unofficial.

As my contribution to the genre I wrote a piece of software that allows me to hunt for asterisms. The minimum number of stars to be in the asterism, the minimum brightness of the stars making up the "design" and the angular size of the asterism can all be pre-selected. The astrometric (positional) and photometric data used by the software was obtained from the UCAC4 catalogue.

Over 3,500 asterisms are included in this collection. Some are unquestionably more visually attractive than others but any one of them will make an interesting diversion from the endlessly recycled objects that alternative publications continue to list!

ANDROMEDA

#	H	M	S	D	M	S	RA DECIMAL	DEC DECIMAL
1	0	15	21.75	47	20	55.27	3.8406	47.3487
2	0	15	54.92	49	16	44.94	3.9788	49.2792
3	1	19	35.63	40	2	51.97	19.8984	40.0478
4	1	38	30.91	44	53	0.92	24.6288	44.8836
5	2	1	38.74	36	13	32.11	30.4114	36.2256
6	2	6	40.43	48	30	33.69	31.6684	48.5094
7	2	16	31.61	45	51	5.68	34.1317	45.8516
8	2	27	47.03	49	41	30.84	36.9460	49.6919
9	23	3	10.5	46	12	11.65	345.7937	46.2032
10	23	15	30	46	34	33.56	348.8750	46.5760
11	23	17	43.31	50	50	25.3	349.4305	50.8404
12	23	19	13.49	47	22	38.47	349.8062	47.3774
13	23	21	25.83	48	12	16.17	350.3576	48.2045
14	23	23	55.62	52	55	26.11	350.9818	52.9239
15	23	24	50.45	46	37	35.99	351.2102	46.6267
16	23	26	6.55	52	41	35.59	351.5273	52.6932
17	23	31	21.75	49	52	49.78	352.8406	49.8805
18	23	39	14.41	47	31	57.85	354.8100	47.5327
19	23	42	14.63	42	27	0.07	355.5610	42.4500
20	23	46	49.92	44	24	0.18	356.7080	44.4001

Nicholson #1 – 15 x 15 arc minutes

ANTLIA

#	H	M	S	D	M	S	RA DECIMAL	DEC DECIMAL
21	9	27	20.21	-38	9	34.46	141.8342	-38.1596
22	9	27	56.6	-37	24	22.82	141.9858	-37.4063
23	9	40	35.45	-36	32	1.04	145.1477	-36.5336
24	10	35	41.38	-39	14	53.16	158.9224	-39.2481

Nicholson #23 – 15 x 15 arc minutes

APUS

#	H	M	S	D	M	S	RA DECIMAL	DEC DECIMAL
25	14	45	28.57	-71	25	51.43	221.3690	-71.4310
26	14	52	17.78	-71	43	16.76	223.0741	-71.7213
27	14	55	16.77	-69	32	3.55	223.8199	-69.5343
28	14	57	50.84	-71	23	34.81	224.4618	-71.3930
29	14	58	58.08	-70	50	58.32	224.7420	-70.8495
30	14	59	28.65	-69	29	14.77	224.8694	-69.4874
31	15	1	13.89	-73	50	2.65	225.3079	-73.8341
32	15	11	8.67	-72	5	46.26	227.7861	-72.0962
33	15	13	18.19	-69	5	52.54	228.3258	-69.0979
34	15	15	20.56	-69	50	6.83	228.8356	-69.8352
35	15	16	49.67	-72	27	23.95	229.2069	-72.4567
36	15	18	40.16	-68	35	59.87	229.6673	-68.6000

37	15	20	22.57	-68	36	53.52	230.0940	-68.6149
38	15	31	18.57	-69	20	56.66	232.8274	-69.3491
39	15	31	31.67	-69	20	41.82	232.8820	-69.3450
40	15	31	50.82	-69	20	3.09	232.9618	-69.3342
41	15	34	0.75	-70	7	22.54	233.5031	-70.1229
42	15	40	36.89	-69	5	34.06	235.1537	-69.0928
43	15	43	58.4	-70	0	40.17	235.9933	-70.0112
44	15	44	10.72	-70	2	2.9	236.0447	-70.0341
45	15	47	28.48	-72	3	15.04	236.8687	-72.0542
46	15	53	57.49	-70	14	3.48	238.4895	-70.2343
47	16	8	9.62	-76	12	16	242.0401	-76.2044
48	17	43	56.34	-71	13	33.2	265.9847	-71.2259

Nicholson #45 – 15 x 15 arc minutes

Nicholson #47 – 15 x 15 arc minutes

AQUILLA

49	18	39	45.18	11	48	45.16	279.9383	11.8125
50	18	40	19.57	9	11	9.17	280.0815	9.1859
51	18	40	58.03	11	38	7.98	280.2418	11.6355
52	18	44	18.62	11	47	59.63	281.0776	11.7999
53	18	46	20.07	8	23	20.69	281.5836	8.3891
54	18	47	14.58	10	58	59.36	281.8107	10.9832
55	18	48	37.26	10	47	23.67	282.1552	10.7899
56	18	52	23.75	14	31	27.78	283.0990	14.5244
57	18	58	45.18	-7	39	55.41	284.6883	-7.6654
58	18	58	48.31	-6	48	5.9	284.7013	-6.8016
59	18	58	54.53	-7	42	14.59	284.7272	-7.7041
60	18	59	16.11	-9	59	47.57	284.8171	-9.9965
61	18	59	25.58	-7	6	34.43	284.8566	-7.1096
62	19	1	35.38	-9	59	51.63	285.3974	-9.9977
63	19	1	41.93	-10	3	16.2	285.4247	-10.0545
64	19	1	56.06	-7	24	50.99	285.4836	-7.4142
65	19	4	52.81	2	45	26.11	286.2200	2.7573
66	19	5	24.33	12	34	33.79	286.3514	12.5761
67	19	5	42.43	-8	34	33.13	286.4268	-8.5759
68	19	10	5.18	-1	24	50.97	287.5216	-1.4142
69	19	11	19.65	3	24	16.97	287.8319	3.4047
70	19	15	4.7	-10	53	20.82	288.7696	-10.8891
71	19	18	17.24	-3	26	51.71	289.5718	-3.4477

72	19	19	12.41	-3	59	10.95	289.8017	-3.9864
73	19	19	48.33	9	28	21.71	289.9514	9.4727
74	19	21	12.14	-11	41	12.2	290.3006	-11.6867
75	19	21	46.73	-10	26	1.1	290.4447	-10.4336
76	19	22	6.47	0	41	7.02	290.5270	-0.6853
77	19	24	55.87	9	1	59.18	291.2328	9.0331
78	19	25	38.1	-2	3	40.02	291.4088	-2.0611
79	19	27	21.94	-7	56	4.09	291.8414	-7.9345
80	19	33	11.88	7	12	8.41	293.2995	7.2023
81	19	34	3.19	11	47	50.4	293.5133	11.7973
82	19	34	36.74	6	15	29.04	293.6531	6.2581
83	19	34	47.1	7	6	21.76	293.6963	7.1060
84	19	35	50.37	6	10	14.46	293.9599	6.1707
85	19	37	9.07	12	51	5.19	294.2878	12.8514
86	19	37	43.53	12	44	10.23	294.4314	12.7362
87	19	37	52.99	12	56	4.12	294.4708	12.9345
88	19	38	22.63	8	54	42.25	294.5943	8.9117
89	19	38	24.66	8	51	25.24	294.6028	8.8570
90	19	38	46.8	9	5	34.46	294.6950	9.0929
91	19	39	16.31	11	21	15.68	294.8179	11.3544
92	19	39	37.78	10	14	15.25	294.9074	10.2376
93	19	40	7.99	11	21	17.58	295.0333	11.3549
94	19	40	45.66	12	28	53.33	295.1903	12.4815
95	19	40	53.61	10	2	46.13	295.2234	10.0461
96	19	41	8.06	13	34	14.59	295.2836	13.5707
97	19	41	12.95	13	31	0.79	295.3040	13.5169
98	19	41	21.87	13	28	51.25	295.3411	13.4809
99	19	41	54.78	13	14	48.4	295.4783	13.2468
100	19	42	2.03	13	17	27.12	295.5084	13.2909
101	19	42	22.19	14	41	20.56	295.5925	14.6890
102	19	43	6.72	12	50	2.22	295.7780	12.8340
103	19	43	30.09	6	28	53.36	295.8754	6.4815
104	19	43	37.48	12	48	1.94	295.9062	12.8005
105	19	43	40.25	12	45	26.94	295.9177	12.7575
106	19	43	43.28	12	43	23.34	295.9303	12.7232
107	19	43	49.49	13	39	39.39	295.9562	13.6609
108	19	43	53.47	4	4	32.41	295.9728	4.0757
109	19	45	28.68	10	13	50.4	296.3695	10.2307
110	19	46	2.39	5	30	16.12	296.5100	5.5045
111	19	46	27.22	4	18	18.06	296.6134	4.3050
112	19	46	35.97	10	23	39.42	296.6499	10.3943
113	19	46	54.86	8	13	20.36	296.7286	8.2223
114	19	48	15.28	4	44	18.36	297.0636	4.7384

115	19	48	52.28	14	24	51.86	297.2178	14.4144
116	19	48	55.48	13	39	47.82	297.2312	13.6633
117	19	48	57.04	14	20	19.12	297.2377	14.3386
118	19	48	57.68	13	38	17.19	297.2403	13.6381
119	19	49	4.51	14	21	50.54	297.2688	14.3640
120	19	53	16.21	9	48	34.43	298.3175	9.8096
121	19	54	44.53	5	7	2.22	298.6856	5.1173
122	19	55	5.92	13	11	38.19	298.7747	13.1939
123	19	55	9.62	4	1	2.21	298.7901	4.0173
124	19	55	31.73	13	30	13.4	298.8822	13.5037
125	19	56	21.64	9	28	5.9	299.0902	9.4683
126	19	56	47.59	9	53	32.22	299.1983	9.8923
127	19	57	4.78	6	43	0.13	299.2699	6.7167
128	19	57	5.26	9	42	40.37	299.2719	9.7112
129	19	57	31.48	4	9	48.76	299.3812	4.1635
130	19	57	59.43	11	30	48.43	299.4976	11.5135
131	20	0	14.96	7	21	35.4	300.0623	7.3598
132	20	0	27.93	5	28	10.86	300.1164	5.4697
133	20	2	25.56	11	55	56.42	300.6065	11.9323
134	20	5	56.67	12	36	23.84	301.4861	12.6066
135	20	6	12.85	13	9	33.2	301.5535	13.1592
136	20	13	6.32	9	57	20.62	303.2763	9.9557

Nicholson #49 – 15 x 15 arc minutes

AQUARIUS

#	H	M	S	D	M	S	RA DECIMAL	DEC DECIMAL
137	20	54	54.43	0	49	46.31	313.7268	0.8295
138	20	56	1.06	-1	19	11.43	314.0044	-1.3198

Nicholson #137 – 15 x 15 arc minutes

ARA

#	H	M	S	D	M	S	RA DECIMAL	DEC DECIMAL
139	16	37	45.66	-46	54	9.04	249.4402	-46.9025
140	16	38	15.1	-48	18	2.35	249.5629	-48.3007
141	16	39	25.64	-54	42	48.72	249.8568	-54.7135
142	16	40	40.85	-47	22	41.48	250.1702	-47.3782
143	16	41	16.82	-47	13	37.14	250.3201	-47.2270
144	16	42	48.5	-54	3	30	250.7021	-54.0583
145	16	43	37.34	-55	15	0.79	250.9056	-55.2502
146	16	43	50.86	-54	17	37.32	250.9619	-54.2937
147	16	45	7.96	-46	31	53.93	251.2832	-46.5316
148	16	45	26.94	-64	40	2.99	251.3622	-64.6675
149	16	46	26.56	-54	3	48.17	251.6107	-54.0634
150	16	47	1.19	-52	19	28.49	251.7550	-52.3246
151	16	48	6.28	-64	18	3.48	252.0262	-64.3010
152	16	48	43.01	-52	18	18.35	252.1792	-52.3051
153	16	48	50.64	-60	3	54.49	252.2110	-60.0651

154	16	49	15.12	-52	27	32.08	252.3130	-52.4589
155	16	49	17.04	-55	9	57.34	252.3210	-55.1659
156	16	50	53.65	-57	38	39.26	252.7235	-57.6442
157	16	51	18.28	-46	50	21.13	252.8262	-46.8392
158	16	52	20.35	-59	52	10.08	253.0848	-59.8695
159	16	52	38.44	-62	47	44.45	253.1602	-62.7957
160	16	52	51.27	-52	53	40.27	253.2136	-52.8945
161	16	52	51.99	-48	48	55.7	253.2166	-48.8155
162	16	53	5.08	-53	13	42.41	253.2712	-53.2284
163	16	53	12.58	-64	8	23.91	253.3024	-64.1400
164	16	53	18.22	-54	8	27.92	253.3259	-54.1411
165	16	53	26.41	-64	5	10.63	253.3600	-64.0863
166	16	54	7.62	-55	5	54.02	253.5317	-55.0983
167	16	55	0.19	-51	58	11.59	253.7508	-51.9699
168	16	56	2.14	-55	56	44.55	254.0089	-55.9457
169	16	56	3.92	-58	49	7.84	254.0163	-58.8188
170	16	56	12.98	-55	53	10.42	254.0541	-55.8862
171	16	56	18.67	-58	47	32.71	254.0778	-58.7924
172	16	56	57.89	-54	51	12.1	254.2412	-54.8534
173	16	59	20.47	-54	52	29.48	254.8353	-54.8749
174	17	0	11.85	-50	51	27.97	255.0494	-50.8578
175	17	1	34.35	-55	54	41.57	255.3931	-55.9115
176	17	2	57.17	-50	10	14.56	255.7382	-50.1707
177	17	5	0.19	-51	22	45.32	256.2508	-51.3793
178	17	5	36.39	-62	18	4.31	256.4016	-62.3012
179	17	7	10.85	-67	18	45.05	256.7952	-67.3125
180	17	7	30.61	-53	47	56.22	256.8776	-53.7990
181	17	7	46.78	-53	45	24.7	256.9449	-53.7569
182	17	9	13.7	-51	8	59.19	257.3071	-51.1498
183	17	9	44.94	-47	16	59.49	257.4372	-47.2832
184	17	10	13.8	-52	23	39.34	257.5575	-52.3943
185	17	12	18	-52	24	47.29	258.0750	-52.4131
186	17	14	7.56	-49	59	33.83	258.5315	-49.9927
187	17	15	25.59	-49	3	45.39	258.8566	-49.0626
188	17	16	15.85	-48	40	57.35	259.0661	-48.6826
189	17	16	37.5	-46	12	10.31	259.1562	-46.2029
190	17	17	23.3	-64	44	22.71	259.3471	-64.7396
191	17	17	44.81	-60	36	7.43	259.4367	-60.6021
192	17	18	3.84	-56	7	12.96	259.5160	-56.1203
193	17	18	16.73	-47	59	51.18	259.5697	-47.9976
194	17	19	7.25	-48	58	11.5	259.7802	-48.9699
195	17	19	34	-54	50	0.11	259.8917	-54.8334
196	17	19	57.06	-54	53	24.52	259.9878	-54.8901

197	17	20	8.66	-46	6	20.8	260.0361	-46.1058
198	17	20	11.56	-55	0	36.16	260.0482	-55.0100
199	17	20	33.89	-46	34	2.28	260.1412	-46.5673
200	17	23	2.78	-45	55	38.85	260.7616	-45.9275
201	17	24	13.19	-50	34	46.12	261.0549	-50.5795
202	17	24	29.2	-58	12	25.15	261.1217	-58.2070
203	17	24	30.54	-50	56	15.73	261.1273	-50.9377
204	17	24	48.56	-65	31	31.78	261.2023	-65.5255
205	17	25	46.38	-47	35	16.06	261.4432	-47.5878
206	17	25	57.19	-46	22	30.43	261.4883	-46.3751
207	17	26	21.03	-46	22	33.46	261.5876	-46.3760
208	17	26	25.38	-51	46	30.56	261.6057	-51.7752
209	17	26	47.89	-50	24	22.43	261.6996	-50.4062
210	17	27	15.31	-46	34	59.54	261.8138	-46.5832
211	17	28	1.19	-50	28	46.87	262.0050	-50.4797
212	17	28	40.92	-65	22	13.24	262.1705	-65.3703
213	17	28	41.44	-48	31	47.93	262.1727	-48.5300
214	17	30	18.21	-47	6	48.33	262.5759	-47.1134
215	17	31	30.48	-56	47	2.02	262.8770	-56.7839
216	17	32	20.23	-61	58	24.22	263.0843	-61.9734
217	17	33	23.76	-51	57	41.64	263.3490	-51.9616
218	17	35	41.5	-62	57	14.55	263.9229	-62.9540
219	17	36	31.77	-46	22	45.59	264.1324	-46.3793
220	17	36	32.74	-49	21	5.1	264.1364	-49.3514
221	17	36	44.47	-46	21	36.44	264.1853	-46.3601
222	17	39	4.5	-54	27	42.94	264.7687	-54.4619
223	17	40	24.82	-47	12	21.64	265.1034	-47.2060
224	17	41	25.32	-48	18	56.68	265.3555	-48.3157
225	17	41	25.39	-50	46	37.41	265.3558	-50.7771
226	17	41	30.9	-48	18	4.61	265.3787	-48.3013
227	17	42	17.53	-48	34	7.95	265.5731	-48.5689
228	17	42	33.28	-66	7	7.08	265.6387	-66.1186
229	17	42	34.11	-46	2	45.92	265.6421	-46.0461
230	17	42	38.77	-66	7	24.21	265.6615	-66.1234
231	17	43	4.15	-47	54	25.9	265.7673	-47.9072
232	17	43	6.39	-51	7	16.06	265.7766	-51.1211
233	17	43	14.53	-47	50	44.17	265.8106	-47.8456
234	17	43	27.74	-51	5	25.91	265.8656	-51.0905
235	17	43	42.06	-48	5	55.97	265.9252	-48.0989
236	17	43	56.02	-47	32	46.14	265.9834	-47.5461
237	17	43	59.28	-49	21	3.66	265.9970	-49.3510
238	17	43	59.59	-47	34	58.47	265.9983	-47.5829
239	17	44	2.13	-49	18	4.64	266.0089	-49.3013

240	17	45	13.83	-48	8	56.62	266.3076	-48.1491
241	17	46	46.27	-57	21	18.11	266.6928	-57.3550
242	17	48	33.17	-51	23	41.09	267.1382	-51.3947
243	17	48	51.08	-46	55	51.81	267.2128	-46.9311
244	17	49	21.74	-47	41	44.58	267.3406	-47.6957
245	17	50	44.75	-51	5	10.05	267.6865	-51.0861
246	17	51	42.17	-45	56	29.52	267.9257	-45.9415
247	17	52	27.19	-45	55	17.54	268.1133	-45.9215
248	17	56	19.68	-46	40	31.32	269.0820	-46.6754
249	17	59	26.87	-47	46	26.44	269.8620	-47.7740
250	17	59	29.56	-47	51	27.01	269.8732	-47.8575
251	17	59	37.88	-47	46	56.46	269.9078	-47.7824
252	17	59	55.65	-47	54	46.72	269.9819	-47.9130

AURIGA

#	H	M	S	D	M	S	RA DECIMAL	DEC DECIMAL
253	4	47	28.24	41	13	13.24	71.8677	41.2203
254	4	52	7.53	42	20	15.97	73.0314	42.3378
255	4	57	52.33	40	54	23.12	74.4680	40.9064
256	5	4	23.98	40	23	33.69	76.0999	40.3927
257	5	8	36.25	40	26	51.67	77.1510	40.4477
258	5	10	57.84	41	0	55.51	77.7410	41.0154
259	5	16	33.81	30	41	1.57	79.1409	30.6838
260	5	19	51.15	41	14	28.57	79.9631	41.2413
261	5	21	39.19	31	28	10.66	80.4133	31.4696
262	5	25	43.27	30	44	39.04	81.4303	30.7442
263	5	29	21.97	33	23	42.23	82.3415	33.3951
264	5	33	53.16	40	42	21.24	83.4715	40.7059
265	5	33	54.75	44	16	5.77	83.4781	44.2683
266	5	35	50.67	36	51	56.75	83.9611	36.8658
267	5	37	11.21	40	0	32.21	84.2967	40.0089
268	5	45	51.29	32	0	56.94	86.4637	32.0158
269	5	48	27.53	42	38	7.01	87.1147	42.6353
270	5	53	10.17	30	42	38.54	88.2924	30.7107
271	5	57	48.55	29	52	12.4	89.4523	29.8701
272	6	2	5.88	30	12	4.31	90.5245	30.2012
273	6	3	37.32	36	3	21.93	90.9055	36.0561
274	6	3	46.57	29	21	2.79	90.9440	29.3508
275	6	6	35.13	28	52	32.88	91.6464	28.8758
276	6	13	20.58	44	28	5.68	93.3358	44.4682
277	6	15	9.07	44	7	28.42	93.7878	44.1246

CAMELOPARDALIS

#	H	M	S	D	M	S	RA DECIMAL	DEC DECIMAL
278	4	48	36.69	55	46	27.62	72.1529	55.7743
279	4	51	6.01	52	14	12.18	72.7750	52.2367

Nicholson #278 – 15 x 15 arc minutes

Nicholson #279 – 15 x 15 arc minutes

CAPRICORNUS

#	H	M	S	D	M	S	RA DECIMAL	DEC DECIMAL
280	20	18	8.83	-24	19	0.73	304.5368	-24.3169

CARINA

#	H	M	S	D	M	S	RA DECIMAL	DEC DECIMAL
281	7	40	46.39	-51	56	25.63	115.1933	-51.9405
282	7	40	53.11	-51	55	37.91	115.2213	-51.9272
283	8	2	28.31	-55	2	12.5	120.6180	-55.0368
284	8	14	16.66	-58	37	35.04	123.5694	-58.6264
285	8	31	46.89	-58	59	13.06	127.9454	-58.9870
286	8	32	2.33	-58	58	47.04	128.0097	-58.9797
287	8	42	37.69	-58	32	26.02	130.6570	-58.5406
288	8	48	36.41	-58	27	6.51	132.1517	-58.4518
289	8	48	38.34	-59	31	27.23	132.1597	-59.5242
290	8	48	40.16	-59	31	28.51	132.1673	-59.5246
291	8	49	24.23	-55	20	37.08	132.3510	-55.3436
292	8	49	44.98	-63	34	49.07	132.4374	-63.5803
293	8	50	41.5	-63	7	28.13	132.6729	-63.1245
294	8	52	14.54	-55	21	19.7	133.0606	-55.3555
295	8	52	22.75	-68	48	59.81	133.0948	-68.8166
296	8	52	50.57	-56	15	5.95	133.2107	-56.2517
297	8	56	6.86	-55	7	50.79	134.0286	-55.1308
298	8	59	13.4	-60	13	30.16	134.8059	-60.2250
299	9	0	23.15	-57	13	58.79	135.0965	-57.2330
300	9	3	3.09	-57	24	23.58	135.7629	-57.4065
301	9	3	14.25	-62	21	23.46	135.8094	-62.3565
302	9	4	18.14	-60	33	57.58	136.0756	-60.5660
303	9	8	25.05	-58	5	51.2	137.1044	-58.0976
304	9	10	12.15	-59	59	17.34	137.5506	-59.9881
305	9	10	36.63	-57	31	48.79	137.6526	-57.5302
306	9	11	55.81	-57	1	18.5	137.9826	-57.0218
307	9	14	3.39	-61	20	4.69	138.5141	-61.3346
308	9	20	53.55	-71	37	19.98	140.2231	-71.6222
309	9	22	30.66	-60	50	11.76	140.6277	-60.8366
310	9	25	5.98	-57	48	12.8	141.2749	-57.8036
311	9	25	19.2	-64	38	44.62	141.3300	-64.6457
312	9	27	1.78	-69	12	58.25	141.7574	-69.2162

313	9	27	49.98	-67	25	36.47	141.9582	-67.4268
314	9	28	25.23	-67	28	0.53	142.1051	-67.4668
315	9	28	56.7	-59	25	49.2	142.2363	-59.4303
316	9	30	19.26	-62	42	7.8	142.5803	-62.7022
317	9	31	15.33	-63	23	32.59	142.8139	-63.3924
318	9	32	17.64	-58	45	56.27	143.0735	-58.7656
319	9	34	7.07	-59	22	47.36	143.5295	-59.3798
320	9	39	29.8	-63	8	43.19	144.8742	-63.1453
321	9	43	2.09	-62	28	15.59	145.7587	-62.4710
322	9	43	18.35	-60	24	11.06	145.8265	-60.4031
323	9	45	22.62	-67	40	27.37	146.3443	-67.6743
324	9	46	1.91	-58	44	37.81	146.5080	-58.7438
325	9	46	42.67	-67	59	46.09	146.6778	-67.9961
326	9	48	30.98	-58	35	29.48	147.1291	-58.5915
327	9	48	32.25	-58	40	34.02	147.1344	-58.6761
328	9	48	44.35	-69	27	14.68	147.1848	-69.4541
329	9	49	53.14	-60	11	17.56	147.4714	-60.1882
330	9	51	11.98	-61	6	24.43	147.7999	-61.1068
331	9	51	37.89	-59	17	17.74	147.9079	-59.2883
332	9	51	52.11	-63	53	4.83	147.9671	-63.8847
333	9	52	10.47	-59	58	36.71	148.0436	-59.9769
334	9	53	2.7	-63	30	46.79	148.2613	-63.5130
335	9	53	3.1	-59	21	50.1	148.2629	-59.3639
336	9	53	18.35	-61	32	1.62	148.3264	-61.5338
337	9	53	22.38	-60	41	21.81	148.3433	-60.6894
338	9	53	41.39	-60	40	41.05	148.4225	-60.6781
339	9	54	35.36	-59	18	3.4	148.6473	-59.3009
340	9	54	38.38	-58	48	26.66	148.6599	-58.8074
341	9	54	46.87	-58	59	59.7	148.6953	-58.9999
342	9	54	57.61	-60	38	26.64	148.7400	-60.6407
343	9	55	5.25	-58	48	49.07	148.7719	-58.8136
344	9	55	41.42	-58	48	22.28	148.9226	-58.8062
345	9	55	46.73	-60	9	34.09	148.9447	-60.1595
346	9	55	50.43	-64	3	20.98	148.9601	-64.0558
347	9	56	1.16	-60	9	46.21	149.0048	-60.1628
348	9	56	24.18	-60	9	18.4	149.1008	-60.1551
349	9	56	54.56	-60	38	38.9	149.2273	-60.6441
350	9	57	11.59	-58	56	2.93	149.2983	-58.9341
351	9	57	13.79	-59	53	41.53	149.3074	-59.8949
352	9	57	27.85	-59	53	31.86	149.3661	-59.8922
353	9	58	21.41	-60	2	51.97	149.5892	-60.0478
354	9	58	25.8	-60	50	44.74	149.6075	-60.8458
355	9	58	33.97	-64	9	0.69	149.6415	-64.1502

356	10	0	9.93	-61	26	13.82	150.0414	-61.4372
357	10	0	11.56	-58	42	41.85	150.0482	-58.7116
358	10	0	12.2	-61	24	32.11	150.0508	-61.4089
359	10	0	33.39	-61	51	48.8	150.1391	-61.8636
360	10	2	45.29	-61	8	57.68	150.6887	-61.1494
361	10	3	36.88	-58	56	45.06	150.9037	-58.9459
362	10	3	40.8	-60	58	2.23	150.9200	-60.9673
363	10	3	49.75	-58	57	10.53	150.9573	-58.9529
364	10	3	54.46	-63	18	4.29	150.9769	-63.3012
365	10	4	43.02	-63	51	1.07	151.1793	-63.8503
366	10	6	31.03	-59	42	2.98	151.6293	-59.7008
367	10	6	52.69	-59	46	54.24	151.7195	-59.7817
368	10	7	37.53	-57	6	1.11	151.9064	-57.1003
369	10	7	40.16	-58	1	26.06	151.9173	-58.0239
370	10	8	3.11	-58	18	19.94	152.0130	-58.3055
371	10	8	37.6	-57	49	36.67	152.1566	-57.8269
372	10	8	43.8	-57	20	13.97	152.1825	-57.3372
373	10	9	49.61	-60	0	43.84	152.4567	-60.0122
374	10	10	34.68	-57	16	41.88	152.6445	-57.2783
375	10	10	39.17	-57	19	33.42	152.6632	-57.3260
376	10	11	8.42	-57	24	30.58	152.7851	-57.4085
377	10	11	26.77	-59	55	5.11	152.8615	-59.9181
378	10	11	30.83	-58	57	48.63	152.8784	-58.9635
379	10	12	4.75	-59	55	50.4	153.0198	-59.9307
380	10	12	10.42	-63	52	26.27	153.0434	-63.8740
381	10	12	16.82	-59	56	42.42	153.0701	-59.9451
382	10	12	19.77	-63	56	45.51	153.0824	-63.9460
383	10	12	28.33	-59	53	34.79	153.1180	-59.8930
384	10	12	54.53	-58	41	11.01	153.2272	-58.6864
385	10	12	58.82	-61	22	59.5	153.2451	-61.3832
386	10	12	59.25	-58	41	21.21	153.2469	-58.6892
387	10	12	59.35	-59	16	29.08	153.2473	-59.2747
388	10	13	3.19	-60	58	5.31	153.2633	-60.9681
389	10	13	6.31	-58	55	59.04	153.2763	-58.9331
390	10	13	7.15	-60	49	13.16	153.2798	-60.8203
391	10	13	7.33	-58	35	53.28	153.2806	-58.5981
392	10	13	20.4	-57	14	25.09	153.3350	-57.2403
393	10	13	21.2	-59	17	12.32	153.3383	-59.2868
394	10	13	23.88	-57	9	59.48	153.3495	-57.1665
395	10	13	28.06	-61	3	56.53	153.3669	-61.0657
396	10	13	28.25	-59	56	19.3	153.3677	-59.9387
397	10	13	31.19	-61	7	19.76	153.3800	-61.1222
398	10	13	38.61	-59	16	10.87	153.4109	-59.2697

399	10	13	39.98	-57	14	3.87	153.4166	-57.2344
400	10	13	51.89	-60	52	1.75	153.4662	-60.8672
401	10	14	3.28	-62	19	15.88	153.5137	-62.3211
402	10	14	11.34	-55	58	51.52	153.5472	-55.9810
403	10	14	32.41	-59	19	38.69	153.6351	-59.3274
404	10	14	33.39	-56	0	19.46	153.6391	-56.0054
405	10	14	36.1	-59	20	31.81	153.6504	-59.3422
406	10	14	36.57	-61	15	10.08	153.6524	-61.2528
407	10	14	49.69	-59	44	57.95	153.7071	-59.7494
408	10	15	46.72	-60	22	47.17	153.9447	-60.3798
409	10	15	49.6	-68	24	39.6	153.9567	-68.4110
410	10	15	52.61	-57	22	30.1	153.9692	-57.3750
411	10	16	5.31	-60	42	42.46	154.0221	-60.7118
412	10	16	16.92	-61	56	21.42	154.0705	-61.9393
413	10	17	26.55	-55	56	44.3	154.3606	-55.9456
414	10	17	52.52	-61	21	41.21	154.4688	-61.3614
415	10	18	4.69	-65	5	22.8	154.5195	-65.0897
416	10	18	13.04	-61	20	53.8	154.5544	-61.3483
417	10	18	50.66	-59	49	48.68	154.7111	-59.8302
418	10	18	55.2	-68	30	54.67	154.7300	-68.5152
419	10	19	0.28	-65	10	49.56	154.7512	-65.1804
420	10	19	6.96	-55	17	23.07	154.7790	-55.2897
421	10	19	12.06	-59	49	13.11	154.8002	-59.8203
422	10	19	58.9	-59	7	35.94	154.9954	-59.1266
423	10	21	4.62	-65	37	28.31	155.2692	-65.6245
424	10	21	8.63	-58	46	13.87	155.2859	-58.7705
425	10	21	8.71	-65	35	51.98	155.2863	-65.5978
426	10	21	10.04	-58	45	30.69	155.2918	-58.7585
427	10	21	15.29	-55	55	29.91	155.3137	-55.9250
428	10	21	20.43	-61	17	12.59	155.3351	-61.2868
429	10	21	32.1	-58	44	29.36	155.3838	-58.7415
430	10	21	36.04	-60	51	10.64	155.4002	-60.8530
431	10	21	53.71	-60	48	29.81	155.4738	-60.8083
432	10	22	6.51	-60	48	23.21	155.5271	-60.8064
433	10	22	10.05	-65	33	36.22	155.5419	-65.5601
434	10	22	12.03	-60	38	45.35	155.5501	-60.6459
435	10	22	23.17	-65	30	39.13	155.5965	-65.5109
436	10	22	35.84	-60	56	7.25	155.6493	-60.9353
437	10	24	0.85	-56	4	38.75	156.0035	-56.0774
438	10	24	0.87	-61	56	21.9	156.0036	-61.9394
439	10	24	3.68	-73	34	21.68	156.0153	-73.5727
440	10	25	0.71	-59	5	50.22	156.2530	-59.0973
441	10	25	30.74	-56	53	57.18	156.3781	-56.8992

442	10	27	41.61	-59	17	4.94	156.9234	-59.2847
443	10	28	7.91	-64	1	35.44	157.0329	-64.0265
444	10	28	16.84	-59	21	0.67	157.0702	-59.3502
445	10	28	17.71	-61	28	25.33	157.0738	-61.4737
446	10	29	9.15	-62	5	1.23	157.2881	-62.0837
447	10	29	15.47	-59	24	10.27	157.3145	-59.4029
448	10	29	21.92	-63	19	50.13	157.3413	-63.3306
449	10	29	43.17	-62	6	25.3	157.4299	-62.1070
450	10	29	45.98	-59	17	19.42	157.4416	-59.2887
451	10	30	30.13	-61	50	46.24	157.6255	-61.8462
452	10	31	10.46	-57	26	0.19	157.7936	-57.4334
453	10	31	30.12	-61	35	33.15	157.8755	-61.5925
454	10	31	49.12	-57	14	25.92	157.9547	-57.2405
455	10	31	59.46	-59	19	41.22	157.9978	-59.3281
456	10	32	1.93	-62	17	43.81	158.0080	-62.2955
457	10	32	41.97	-57	39	9.67	158.1749	-57.6527
458	10	33	17.61	-57	29	53.72	158.3234	-57.4983
459	10	33	23.19	-61	10	19.66	158.3466	-61.1721
460	10	33	31.27	-57	32	33.57	158.3803	-57.5427
461	10	33	37.04	-60	37	16.13	158.4043	-60.6211
462	10	33	38.57	-57	36	23.12	158.4107	-57.6064
463	10	33	48.3	-60	41	17.37	158.4512	-60.6882
464	10	33	52.5	-59	27	2.42	158.4687	-59.4507
465	10	34	2.68	-63	53	28.82	158.5112	-63.8913
466	10	34	8.84	-59	28	54.01	158.5369	-59.4817
467	10	34	35.25	-72	38	57.54	158.6469	-72.6493
468	10	34	53.75	-72	40	23.94	158.7240	-72.6733
469	10	35	0.7	-60	54	37.73	158.7529	-60.9105
470	10	35	5.36	-72	36	37.64	158.7723	-72.6105
471	10	35	16.31	-67	25	13.03	158.8180	-67.4203
472	10	35	21.25	-61	8	54.43	158.8385	-61.1485
473	10	36	1.3	-60	53	38.42	159.0054	-60.8940
474	10	36	12.11	-57	38	39.12	159.0505	-57.6442
475	10	36	16.09	-60	54	27.65	159.0671	-60.9077
476	10	37	55.65	-61	6	29.46	159.4819	-61.1082
477	10	38	9.21	-61	18	14.12	159.5384	-61.3039
478	10	38	18.12	-60	32	56.12	159.5755	-60.5489
479	10	38	19.26	-61	17	23.37	159.5803	-61.2898
480	10	38	45.72	-61	14	7.65	159.6905	-61.2355
481	10	38	47	-63	8	43.47	159.6958	-63.1454
482	10	39	12.82	-60	47	55.74	159.8034	-60.7988
483	10	39	22.97	-60	49	19.05	159.8457	-60.8220
484	10	39	24.69	-56	30	17.33	159.8529	-56.5048

485	10	39	29.64	-60	21	43.64	159.8735	-60.3621
486	10	39	47.72	-56	25	20.93	159.9488	-56.4225
487	10	39	53.64	-60	23	32.7	159.9735	-60.3924
488	10	40	34.53	-56	18	23.81	160.1439	-56.3066
489	10	40	36.03	-60	26	36	160.1501	-60.4433
490	10	40	40.93	-56	33	6.33	160.1705	-56.5518
491	10	40	43.96	-56	21	29.94	160.1832	-56.3583
492	10	41	0.17	-57	36	2.97	160.2507	-57.6008
493	10	41	25.31	-61	38	36.64	160.3554	-61.6435
494	10	41	25.63	-57	36	5.13	160.3568	-57.6014
495	10	41	35.32	-69	42	41.86	160.3971	-69.7116
496	10	41	39.01	-61	40	41.16	160.4125	-61.6781
497	10	41	40.84	-60	36	37.78	160.4202	-60.6105
498	10	42	0.41	-61	44	44.1	160.5017	-61.7456
499	10	42	28.32	-56	50	41.27	160.6180	-56.8448
500	10	42	56.7	-56	32	12.43	160.7363	-56.5368
501	10	43	5.39	-61	40	58.63	160.7725	-61.6830
502	10	43	10.26	-56	28	8.93	160.7927	-56.4691
503	10	43	15.91	-58	17	17.12	160.8163	-58.2881
504	10	43	18.47	-61	45	25.54	160.8269	-61.7571
505	10	43	54.66	-58	5	16.03	160.9777	-58.0878
506	10	43	59.01	-57	21	27.36	160.9959	-57.3576
507	10	44	5.93	-58	6	38.42	161.0247	-58.1107
508	10	44	14.87	-57	59	5.07	161.0620	-57.9847
509	10	44	20.05	-58	3	53.46	161.0835	-58.0649
510	10	44	20.63	-62	4	12.54	161.0859	-62.0701
511	10	44	26.07	-57	59	12.57	161.1086	-57.9868
512	10	44	29.47	-62	5	4.05	161.1228	-62.0845
513	10	45	4.12	-58	8	55.52	161.2672	-58.1488
514	10	45	5.81	-61	37	38.62	161.2742	-61.6274
515	10	45	12.12	-58	11	32.04	161.3005	-58.1922
516	10	46	12.61	-58	38	48.79	161.5525	-58.6469
517	10	46	19.37	-61	39	22.86	161.5807	-61.6564
518	10	46	33.44	-58	35	52.85	161.6393	-58.5980
519	10	47	32.15	-61	2	0.09	161.8839	-61.0334
520	10	48	2.74	-60	44	43.64	162.0114	-60.7455
521	10	48	6.33	-61	40	37.58	162.0264	-61.6771
522	10	48	11.21	-60	41	25.82	162.0467	-60.6905
523	10	48	26.7	-61	57	23.52	162.1112	-61.9565
524	10	48	32.84	-60	38	12.36	162.1369	-60.6368
525	10	48	33.03	-61	50	49.57	162.1376	-61.8471
526	10	48	35.28	-60	35	21.39	162.1470	-60.5893
527	10	48	41.96	-63	49	58.83	162.1748	-63.8330

528	10	48	43.53	-59	26	52.8	162.1814	-59.4480
529	10	48	45.31	-63	49	40.94	162.1888	-63.8280
530	10	48	46.52	-60	35	39.87	162.1939	-60.5944
531	10	48	54.17	-61	29	48.88	162.2257	-61.4969
532	10	48	58.63	-60	36	13.78	162.2443	-60.6038
533	10	50	47.75	-63	42	52.13	162.6989	-63.7145
534	10	50	53.21	-66	19	13.62	162.7217	-66.3204
535	10	50	57.08	-63	38	2.02	162.7378	-63.6339
536	10	51	15.95	-57	36	59.87	162.8164	-57.6166
537	10	52	3.84	-60	48	14.37	163.0160	-60.8040
538	10	52	14.73	-57	23	9.63	163.0614	-57.3860
539	10	52	26.5	-56	58	36.52	163.1104	-56.9768
540	10	53	0.92	-57	7	10.99	163.2538	-57.1197
541	10	53	55.23	-58	47	7.79	163.4801	-58.7855
542	10	54	19.12	-56	53	26.03	163.5797	-56.8906
543	10	54	51.56	-56	51	24.65	163.7148	-56.8568
544	10	55	24.47	-68	21	26.33	163.8520	-68.3573
545	10	56	58	-56	53	29.24	164.2417	-56.8915
546	10	57	10.97	-68	44	35.34	164.2957	-68.7432
547	10	57	31.09	-58	26	20.29	164.3796	-58.4390
548	10	57	53.88	-56	7	29.27	164.4745	-56.1248
549	10	58	30.57	-56	49	7.96	164.6274	-56.8189
550	10	59	16.83	-57	51	21.6	164.8201	-57.8560
551	10	59	24.15	-62	37	8.97	164.8506	-62.6192
552	10	59	53.79	-56	4	52.26	164.9741	-56.0812
553	10	59	57.19	-56	45	52.66	164.9883	-56.7646
554	10	59	59.1	-56	0	44.52	164.9963	-56.0124
555	11	0	3.12	-57	35	23.69	165.0130	-57.5899
556	11	0	3.15	-56	2	53.39	165.0131	-56.0482
557	11	0	3.57	-57	35	16.24	165.0149	-57.5878
558	11	0	6.09	-56	42	23.11	165.0254	-56.7064
559	11	0	8.37	-56	1	52.93	165.0349	-56.0314
560	11	0	13.72	-58	12	57.6	165.0572	-58.2160
561	11	0	21.26	-57	31	45.95	165.0886	-57.5294
562	11	0	33.34	-56	37	12.51	165.1389	-56.6201
563	11	0	34.65	-57	41	50.3	165.1444	-57.6973
564	11	0	36.14	-72	59	1.12	165.1506	-72.9836
565	11	0	51.08	-56	9	13.05	165.2128	-56.1536
566	11	1	4.22	-56	12	21.81	165.2676	-56.2061
567	11	1	23.86	-58	1	41.96	165.3494	-58.0283
568	11	1	28.66	-61	58	58.02	165.3694	-61.9828
569	11	1	29.54	-58	1	55.91	165.3731	-58.0322
570	11	1	42.84	-58	0	8.61	165.4285	-58.0024

571	11	1	47.07	-62	16	45.8	165.4461	-62.2794
572	11	1	49.92	-56	31	52.18	165.4580	-56.5312
573	11	1	51.21	-58	0	19.95	165.4634	-58.0055
574	11	2	0.05	-56	27	4.91	165.5002	-56.4514
575	11	2	13.08	-61	52	42.05	165.5545	-61.8783
576	11	2	15.01	-57	23	2.33	165.5625	-57.3840
577	11	2	45.69	-68	25	58.21	165.6904	-68.4328
578	11	2	55.32	-61	52	31.71	165.7305	-61.8755
579	11	3	5.92	-61	53	27.2	165.7747	-61.8909
580	11	3	11.5	-58	8	18.53	165.7979	-58.1385
581	11	3	12.74	-61	54	40.88	165.8031	-61.9114
582	11	3	31.56	-58	5	41.34	165.8815	-58.0948
583	11	3	33.41	-59	12	15.54	165.8892	-59.2043
584	11	3	38.68	-57	33	45.44	165.9112	-57.5626
585	11	3	48.76	-58	9	48.92	165.9532	-58.1636
586	11	3	54.26	-57	31	59.67	165.9761	-57.5332
587	11	4	12.84	-55	1	58.11	166.0535	-55.0328
588	11	4	19.06	-69	44	9.08	166.0794	-69.7359
589	11	4	57.22	-55	41	8.73	166.2384	-55.6858
590	11	5	2.36	-60	27	54.6	166.2598	-60.4652
591	11	5	13.43	-57	14	7.15	166.3060	-57.2353
592	11	5	13.46	-56	13	48.52	166.3061	-56.2301
593	11	5	25.34	-64	38	17.56	166.3556	-64.6382
594	11	6	3.09	-55	26	55.77	166.5129	-55.4488
595	11	6	14.88	-58	4	41	166.5620	-58.0781
596	11	6	36.71	-57	13	40.46	166.6529	-57.2279
597	11	7	51.76	-56	39	37.37	166.9656	-56.6604
598	11	8	35.53	-67	6	9.61	167.1480	-67.1027
599	11	8	56.15	-71	3	30.93	167.2339	-71.0586
600	11	9	1.62	-56	9	53.03	167.2568	-56.1647
601	11	9	13.35	-57	16	52.94	167.3056	-57.2814
602	11	10	12.9	-55	16	4.42	167.5537	-55.2679
603	11	11	9.75	-57	16	40.93	167.7906	-57.2780
604	11	11	42.73	-58	12	36.91	167.9280	-58.2103
605	11	12	11.3	-58	8	17.29	168.0471	-58.1381
606	11	13	23.27	-57	1	20.81	168.3469	-57.0224
607	11	13	31.12	-57	3	6.38	168.3797	-57.0518
608	11	14	23.58	-63	23	41.15	168.5983	-63.3948
609	11	15	16.09	-58	10	52.88	168.8170	-58.1814
610	11	15	17.05	-59	31	30.2	168.8210	-59.5251
611	11	15	17.76	-56	13	45.76	168.8240	-56.2294
612	11	15	19.18	-58	8	59.66	168.8299	-58.1499
613	11	15	33.67	-56	29	27.38	168.8903	-56.4909

614	11	16	18.11	-72	3	0.78	169.0755	-72.0502
615	11	16	24.44	-59	41	36.26	169.1018	-59.6934
616	11	16	26.23	-72	1	23.54	169.1093	-72.0232
617	11	17	15.7	-56	8	7.87	169.3154	-56.1355
618	11	18	14.18	-61	54	7.87	169.5591	-61.9022
619	11	18	48.09	-60	53	57.44	169.7004	-60.8993
620	11	19	45.93	-57	12	30.83	169.9414	-57.2086
621	11	20	55.77	-56	58	51.9	170.2324	-56.9811
622	11	20	59.82	-61	5	7.52	170.2493	-61.0854
623	11	21	15.09	-60	32	44.62	170.3129	-60.5457
624	11	22	1.45	-60	58	11.82	170.5060	-60.9700
625	11	22	4.05	-57	14	7.91	170.5169	-57.2355
626	11	22	23.98	-60	58	46.51	170.5999	-60.9796
627	11	22	28.79	-61	16	44.24	170.6199	-61.2790
628	11	22	41.77	-60	49	32.01	170.6740	-60.8256
629	11	23	7.9	-60	56	2.25	170.7829	-60.9340
630	11	23	10.63	-61	42	31.04	170.7943	-61.7086
631	11	24	23.58	-61	44	20.57	171.0983	-61.7390
632	11	24	34.51	-61	31	49.9	171.1438	-61.5305
633	11	24	56.44	-61	39	45.41	171.2352	-61.6626
634	11	25	7.44	-61	31	38.76	171.2810	-61.5274
635	11	25	30.86	-67	46	43.08	171.3786	-67.7786

Nicholson #634 – 15 x 15 arc minutes

CASSIOPEIA

#	H	M	S	D	M	S	RA DECIMAL	DEC DECIMAL
636	0	2	46.01	58	30	14.68	0.6917	58.5041
637	0	8	48.45	54	37	26.48	2.2019	54.6240
638	0	14	44.99	53	54	9.07	3.6875	53.9025
639	0	15	4.18	54	37	36.99	3.7674	54.6269
640	0	18	50.37	54	25	50.98	4.7099	54.4308
641	0	22	59.16	54	17	26.3	5.7465	54.2906
642	0	26	56.88	50	42	3.73	6.7370	50.7010
643	0	31	37.6	50	54	3.88	7.9067	50.9011
644	0	34	30.08	60	6	33.63	8.6253	60.1093
645	0	38	16.26	57	37	29.97	9.5678	57.6250
646	0	38	20.9	62	49	28.28	9.5871	62.8245
647	0	47	39.3	71	1	53.6	11.9137	71.0316
648	0	48	11.19	59	44	11.24	12.0466	59.7365
649	0	52	35.81	52	56	12.61	13.1492	52.9368
650	0	54	2.68	54	35	39.53	13.5112	54.5943
651	0	55	10.12	58	27	33.88	13.7921	58.4594
652	0	56	19.23	52	52	20.02	14.0801	52.8722
653	1	0	20.78	55	22	1.59	15.0866	55.3671
654	1	6	36.99	70	54	25.65	16.6541	70.9071
655	1	8	55.47	62	58	3.43	17.2311	62.9676
656	1	14	6.09	57	7	0.36	18.5254	57.1168
657	1	14	30.29	60	53	28.8	18.6262	60.8913
658	1	21	18.72	56	44	18.39	20.3280	56.7384
659	1	22	29.07	56	24	46.55	20.6211	56.4129
660	1	23	38.81	55	40	0.53	20.9117	55.6668
661	1	23	56.84	62	11	5.79	20.9868	62.1849
662	1	24	57.18	57	20	12.97	21.2382	57.3369
663	1	27	27.08	57	22	11.1	21.8628	57.3698
664	1	27	53.31	55	42	47.29	21.9721	55.7131
665	1	28	42.77	73	57	5.61	22.1782	73.9516
666	1	36	14.05	54	0	31.11	24.0585	54.0086
667	1	39	36.67	63	26	44.56	24.9028	63.4457
668	1	45	14.99	57	9	8.1	26.3125	57.1523
669	1	48	24.61	56	3	44.85	27.1025	56.0625
670	1	52	12.71	58	33	41.24	28.0530	58.5615
671	2	8	5.19	58	35	46.28	32.0216	58.5962
672	2	24	25.67	62	59	43.73	36.1070	62.9955
673	2	37	58.16	62	55	33.58	39.4923	62.9260
674	22	46	56.01	54	57	7.45	341.7334	54.9521

675	22	53	26.86	54	0	4.59	343.3619	54.0013
676	22	53	56.52	53	19	21.5	343.4855	53.3226
677	22	54	5.98	53	18	35.64	343.5249	53.3099
678	22	59	37.02	54	46	55.42	344.9042	54.7821
679	23	6	31.93	52	45	1.04	346.6330	52.7503
680	23	9	59.61	53	37	42.28	347.4984	53.6284
681	23	11	26.66	53	1	53.32	347.8611	53.0315
682	23	14	21.18	52	30	29.89	348.5883	52.5083
683	23	16	59.35	54	59	1.6	349.2473	54.9838
684	23	17	7.34	55	29	7.11	349.2806	55.4853
685	23	19	5.43	55	48	23.12	349.7726	55.8064
686	23	26	38.09	54	30	9.1	351.6587	54.5025
687	23	31	56.08	54	1	14	352.9837	54.0206
688	23	34	24.35	54	0	44.25	353.6014	54.0123
689	23	35	39.01	55	50	58.31	353.9126	55.8495
690	23	42	51.94	61	28	46.25	355.7164	61.4795
691	23	44	31.42	51	34	19.36	356.1309	51.5720
692	23	48	9.63	56	5	38.01	357.0401	56.0939
693	23	50	28.46	64	40	33.19	357.6186	64.6759
694	23	53	22.79	58	58	25.45	358.3450	58.9737
695	23	57	35.72	62	8	21.45	359.3988	62.1393
696	23	59	59.98	56	17	22.9	359.9999	56.2897

Nicholson #696 – 15 x 15 arc minutes

CENTAURUS

#	H	M	S	D	M	S	RA DECIMAL	DEC DECIMAL
697	11	10	48.8	-52	20	37.99	167.7033	-52.3439
698	11	11	48.77	-52	48	43.01	167.9532	-52.8119
699	11	13	24.09	-52	23	51.87	168.3504	-52.3977
700	11	17	55.51	-52	52	3.85	169.4813	-52.8677
701	11	18	54.24	-54	4	52.18	169.7260	-54.0812
702	11	19	8.52	-50	42	15.86	169.7855	-50.7044
703	11	19	40.03	-54	5	49.59	169.9168	-54.0971
704	11	21	46.43	-55	42	18.1	170.4435	-55.7050
705	11	22	24.14	-60	27	1.5	170.6006	-60.4504
706	11	22	50.55	-55	2	56.02	170.7106	-55.0489
707	11	22	57.83	-59	20	59.73	170.7410	-59.3499
708	11	23	20.25	-59	47	55.02	170.8344	-59.7986
709	11	23	44.09	-59	6	30.59	170.9337	-59.1085
710	11	23	48.76	-57	51	39.66	170.9532	-57.8610
711	11	23	52.32	-59	3	59.51	170.9680	-59.0665
712	11	24	8.31	-53	53	30.68	171.0346	-53.8919
713	11	24	21.86	-55	56	44.87	171.0911	-55.9458
714	11	24	30.06	-59	12	24.04	171.1253	-59.2067
715	11	24	38.76	-59	15	17.96	171.1615	-59.2550
716	11	25	5.31	-59	10	14.91	171.2721	-59.1708
717	11	25	40.69	-57	37	20.18	171.4195	-57.6223
718	11	25	41.14	-55	53	48.94	171.4214	-55.8969
719	11	25	44.36	-57	36	58.23	171.4348	-57.6162
720	11	25	53.89	-55	24	25.75	171.4746	-55.4072
721	11	26	2.4	-55	54	19.86	171.5100	-55.9055
722	11	26	18.64	-59	21	7.71	171.5777	-59.3521
723	11	26	26.58	-56	54	53.84	171.6107	-56.9150
724	11	27	16.82	-59	0	36.46	171.8201	-59.0101
725	11	27	47.83	-61	30	23.94	171.9493	-61.5066
726	11	27	57.74	-57	41	41.2	171.9906	-57.6948
727	11	28	3.88	-61	27	5.8	172.0162	-61.4516
728	11	28	13.27	-57	44	19.49	172.0553	-57.7387
729	11	29	23.18	-55	56	56.37	172.3466	-55.9490
730	11	29	30.28	-62	22	29.53	172.3762	-62.3749
731	11	29	56.34	-56	54	34.78	172.4847	-56.9097
732	11	29	58.13	-55	17	37.44	172.4922	-55.2937
733	11	31	13.76	-57	10	12.27	172.8073	-57.1701
734	11	31	42.13	-49	30	16.58	172.9256	-49.5046
735	11	32	13.39	-57	46	6.35	173.0558	-57.7684

736	11	32	16.58	-62	32	51.83	173.0691	-62.5477
737	11	32	32.17	-56	9	51.35	173.1340	-56.1643
738	11	32	47.79	-61	57	42.48	173.1991	-61.9618
739	11	32	52.86	-55	2	12.28	173.2202	-55.0367
740	11	33	4.83	-59	22	54.38	173.2701	-59.3818
741	11	33	15.34	-57	31	40.16	173.3139	-57.5278
742	11	33	58.15	-59	45	32.29	173.4923	-59.7590
743	11	34	2.92	-59	4	8.47	173.5122	-59.0690
744	11	34	31.1	-58	43	30.12	173.6296	-58.7250
745	11	34	40.35	-56	20	6.15	173.6681	-56.3350
746	11	35	6.81	-57	40	51.52	173.7784	-57.6810
747	11	35	13.75	-56	23	27.95	173.8073	-56.3911
748	11	35	57.48	-64	4	34.19	173.9895	-64.0762
749	11	36	4.36	-59	56	7.68	174.0182	-59.9355
750	11	36	43.11	-59	26	25.46	174.1796	-59.4404
751	11	36	48.45	-57	40	56.64	174.2019	-57.6824
752	11	36	48.47	-59	29	29.43	174.2020	-59.4915
753	11	36	53.58	-57	40	42.63	174.2233	-57.6785
754	11	36	58.12	-61	6	34.27	174.2422	-61.1095
755	11	37	20.21	-56	25	26.09	174.3342	-56.4239
756	11	37	37.05	-60	45	0.25	174.4044	-60.7501
757	11	37	58.76	-59	42	51.47	174.4948	-59.7143
758	11	38	5.21	-58	43	13.9	174.5217	-58.7205
759	11	38	25.97	-59	45	32.08	174.6082	-59.7589
760	11	39	10.21	-60	27	3.68	174.7925	-60.4510
761	11	39	10.72	-57	44	30.9	174.7947	-57.7419
762	11	39	15.27	-60	26	11.52	174.8136	-60.4365
763	11	39	21.17	-57	42	53.3	174.8382	-57.7148
764	11	39	22.59	-60	27	44.65	174.8441	-60.4624
765	11	39	35.55	-57	43	42.73	174.8981	-57.7285
766	11	39	48.13	-60	26	14.38	174.9505	-60.4373
767	11	39	56.12	-61	52	4.78	174.9838	-61.8680
768	11	40	3.5	-56	32	26.94	175.0146	-56.5408
769	11	40	8.13	-53	56	49.96	175.0339	-53.9472
770	11	40	35.84	-60	25	0.54	175.1493	-60.4168
771	11	41	13.66	-57	39	34.94	175.3069	-57.6597
772	11	41	43.47	-60	16	30.49	175.4311	-60.2751
773	11	42	2.07	-61	41	13.61	175.5086	-61.6871
774	11	42	19.05	-57	33	55.17	175.5794	-57.5653
775	11	42	47.64	-37	33	2.99	175.6985	-37.5508
776	11	42	56.07	-60	13	1.83	175.7336	-60.2172
777	11	43	19.77	-59	0	27.23	175.8324	-59.0076
778	11	43	31.43	-57	59	13.14	175.8810	-57.9870

779	11	43	44.85	-59	48	33.5	175.9369	-59.8093
780	11	43	47.1	-50	53	35.88	175.9463	-50.8933
781	11	44	15.59	-61	42	59.07	176.0650	-61.7164
782	11	44	28.57	-60	23	43.68	176.1190	-60.3955
783	11	44	40.15	-51	56	36.37	176.1673	-51.9434
784	11	44	48.56	-60	23	51.08	176.2023	-60.3975
785	11	45	20.95	-63	33	10.24	176.3373	-63.5528
786	11	45	22.63	-61	47	51.32	176.3443	-61.7976
787	11	45	48.22	-63	34	29.03	176.4509	-63.5747
788	11	46	13.04	-57	4	4.46	176.5543	-57.0679
789	11	46	24.07	-55	12	42.32	176.6003	-55.2118
790	11	47	33.92	-60	11	36.64	176.8913	-60.1935
791	11	47	41.26	-59	36	35.09	176.9219	-59.6097
792	11	47	58.13	-59	36	3.61	176.9922	-59.6010
793	11	48	6.81	-58	19	14.76	177.0284	-58.3208
794	11	48	23.21	-61	37	14.4	177.0967	-61.6207
795	11	48	56.9	-57	45	31.66	177.2371	-57.7588
796	11	49	2.65	-59	53	54.85	177.2611	-59.8986
797	11	49	23.26	-60	37	56.95	177.3469	-60.6325
798	11	49	24.37	-53	43	40.22	177.3515	-53.7278
799	11	49	39.44	-60	36	18.27	177.4143	-60.6051
800	11	49	40.83	-58	18	0.32	177.4201	-58.3001
801	11	49	45.33	-60	34	25.44	177.4389	-60.5737
802	11	50	24.7	-49	38	37.55	177.6029	-49.6438
803	11	51	21.4	-54	50	42.38	177.8392	-54.8451
804	11	51	51.1	-58	2	33.9	177.9620	-58.0428
805	11	54	44.99	-56	19	49.02	178.6875	-56.3303
806	11	54	58.7	-54	38	2.84	178.7446	-54.6341
807	11	55	5.14	-54	39	7.74	178.7714	-54.6522
808	11	55	11.73	-53	49	57.49	178.7989	-53.8326
809	11	55	46.68	-55	33	14.63	178.9445	-55.5541
810	11	56	14.3	-45	46	23.14	179.0596	-45.7731
811	11	57	4.59	-55	14	46.28	179.2691	-55.2462
812	11	58	23.63	-55	36	3.03	179.5984	-55.6008
813	11	58	31.03	-55	37	11.99	179.6293	-55.6200
814	11	58	41.02	-55	39	19.87	179.6709	-55.6555
815	11	58	58.4	-58	0	15.44	179.7433	-58.0043
816	11	59	32.96	-50	27	12.7	179.8873	-50.4535
817	12	0	31.28	-56	17	38.53	180.1303	-56.2940
818	12	1	10.23	-55	41	43.32	180.2926	-55.6954
819	12	1	21.92	-55	40	33.89	180.3413	-55.6761
820	12	1	33.78	-54	44	36.62	180.3907	-54.7435
821	12	3	55.47	-54	38	57.5	180.9811	-54.6493

822	12	4	35.34	-58	41	26.27	181.1472	-58.6906
823	12	4	37.55	-58	41	59.69	181.1565	-58.6999
824	12	4	54.45	-58	43	58.18	181.2269	-58.7328
825	12	4	56.99	-53	55	38.41	181.2375	-53.9273
826	12	8	10.83	-50	0	5.58	182.0451	-50.0016
827	12	9	30.9	-54	8	12.65	182.3787	-54.1368
828	12	11	4.66	-56	16	53.57	182.7694	-56.2815
829	12	18	10.01	-56	29	35.05	184.5417	-56.4931
830	12	20	20.44	-53	8	2.54	185.0852	-53.1340
831	12	20	51.14	-55	48	11.53	185.2131	-55.8032
832	12	22	29.85	-50	23	55.32	185.6244	-50.3987
833	12	22	34.88	-55	9	58.26	185.6453	-55.1662
834	12	41	26.83	-53	48	45.83	190.3618	-53.8127
835	12	44	58.11	-42	28	13.4	191.2421	-42.4704
836	12	51	58.59	-55	21	42.95	192.9941	-55.3619
837	12	52	27.97	-59	4	27.67	193.1165	-59.0744
838	12	52	37.98	-57	22	22.44	193.1583	-57.3729
839	12	52	43.49	-59	6	11.62	193.1812	-59.1032
840	12	54	3.79	-51	49	55.75	193.5158	-51.8322
841	12	58	31.89	-58	15	22.43	194.6329	-58.2562
842	13	3	2.99	-59	57	37.9	195.7624	-59.9605
843	13	3	19.59	-56	5	46.34	195.8316	-56.0962
844	13	4	58.23	-60	6	43.08	196.2426	-60.1120
845	13	7	7.09	-57	29	31.37	196.7795	-57.4920
846	13	8	28.28	-61	12	9.41	197.1178	-61.2026
847	13	8	43.06	-61	11	10.98	197.1794	-61.1864
848	13	9	19.88	-55	35	58.46	197.3328	-55.5996
849	13	9	24.67	-60	14	57.95	197.3528	-60.2494
850	13	10	16.73	-57	0	32.22	197.5697	-57.0090
851	13	10	32.73	-60	51	23.88	197.6364	-60.8566
852	13	11	2.11	-60	7	21.66	197.7588	-60.1227
853	13	11	3.39	-64	28	5.3	197.7641	-64.4681
854	13	11	5.41	-60	9	50.43	197.7725	-60.1640
855	13	11	7.8	-64	20	37.01	197.7825	-64.3436
856	13	11	54.83	-62	11	11.05	197.9785	-62.1864
857	13	11	59.06	-64	34	2.47	197.9961	-64.5674
858	13	14	3.95	-61	42	21.88	198.5165	-61.7061
859	13	15	37.45	-64	33	55.69	198.9060	-64.5655
860	13	15	59.44	-61	55	49.79	198.9977	-61.9305
861	13	16	33.6	-61	32	18.58	199.1400	-61.5385
862	13	17	21.5	-63	56	3.05	199.3396	-63.9342
863	13	19	39.91	-60	59	17.74	199.9163	-60.9883
864	13	20	56.09	-58	32	31.76	200.2337	-58.5422

865	13	21	32.21	-60	7	40.42	200.3842	-60.1279
866	13	21	46.59	-60	52	32.81	200.4441	-60.8758
867	13	25	9.65	-61	5	24.28	201.2902	-61.0901
868	13	25	25.49	-64	25	0.13	201.3562	-64.4167
869	13	25	25.82	-46	39	19.2	201.3576	-46.6553
870	13	26	7.92	-61	37	58.41	201.5330	-61.6329
871	13	26	15.27	-58	19	43.42	201.5636	-58.3287
872	13	26	20.95	-61	36	10.54	201.5873	-61.6029
873	13	31	48.91	-50	3	56.18	202.9538	-50.0656
874	13	32	48.22	-61	45	39.87	203.2009	-61.7611
875	13	33	41.9	-63	47	13	203.4246	-63.7869
876	13	33	49.16	-61	37	27.17	203.4548	-61.6242
877	13	34	18.12	-63	58	54.85	203.5755	-63.9819
878	13	34	35.16	-53	14	57.56	203.6465	-53.2493
879	13	35	45.91	-63	20	27.03	203.9413	-63.3408
880	13	36	14.95	-61	26	42.1	204.0623	-61.4450
881	13	38	13.14	-58	21	42.22	204.5547	-58.3617
882	13	38	40.36	-62	42	51.33	204.6682	-62.7143
883	13	39	10.16	-52	41	6.97	204.7923	-52.6853
884	13	39	50.85	-54	31	33.62	204.9619	-54.5260
885	13	40	32.65	-60	33	59.07	205.1360	-60.5664
886	13	40	34.75	-59	52	40.53	205.1448	-59.8779
887	13	40	54.51	-60	35	56.04	205.2271	-60.5989
888	13	43	23.21	-60	57	20.15	205.8467	-60.9556
889	13	43	32.2	-45	34	35.26	205.8842	-45.5765
890	13	43	34.52	-58	32	31.1	205.8938	-58.5420
891	13	44	11.09	-47	4	0.41	206.0462	-47.0668
892	13	45	29.25	-61	16	51.71	206.3719	-61.2810
893	13	45	51.96	-49	6	37.78	206.4665	-49.1105
894	13	46	28.33	-61	20	57.66	206.6180	-61.3494
895	13	47	5.97	-60	43	48.91	206.7749	-60.7303
896	13	47	32.48	-60	41	54.13	206.8853	-60.6984
897	13	48	16	-58	13	28.24	207.0667	-58.2245
898	13	48	32.97	-58	36	10.35	207.1374	-58.6029
899	13	48	58.69	-58	39	47.75	207.2446	-58.6633
900	13	49	15.84	-60	48	56.79	207.3160	-60.8158
901	13	52	1.82	-58	29	34.98	208.0076	-58.4931
902	13	52	2.1	-60	52	33.37	208.0087	-60.8759
903	13	52	3.59	-58	31	10.05	208.0150	-58.5195
904	13	52	11.8	-60	52	17.38	208.0492	-60.8715
905	13	52	27.72	-58	30	43.47	208.1155	-58.5121
906	13	53	55.16	-62	36	39.89	208.4798	-62.6111
907	13	53	55.99	-57	51	35.5	208.4833	-57.8599

908	13	54	16.91	-62	36	53.41	208.5705	-62.6148
909	13	54	45.81	-58	9	25.05	208.6909	-58.1570
910	13	55	0.77	-62	29	20.41	208.7532	-62.4890
911	13	55	7.27	-57	40	18.62	208.7803	-57.6718
912	13	55	31.89	-56	43	10.12	208.8829	-56.7195
913	13	55	48.97	-62	22	38.55	208.9540	-62.3774
914	13	57	1.59	-62	47	48.2	209.2566	-62.7967
915	13	57	6.73	-53	23	41.72	209.2780	-53.3949
916	13	57	11.34	-62	47	58.78	209.2973	-62.7997
917	13	57	38.08	-62	48	38.58	209.4087	-62.8107
918	13	57	54.52	-45	26	18.95	209.4772	-45.4386
919	13	59	37.09	-63	9	28.95	209.9045	-63.1580
920	13	59	59.72	-50	53	26.24	209.9988	-50.8906
921	14	1	4.19	-62	43	13.97	210.2675	-62.7205
922	14	1	55.31	-60	47	35.27	210.4805	-60.7931
923	14	2	18.94	-61	21	2.5	210.5789	-61.3507
924	14	2	43.18	-56	8	58.95	210.6799	-56.1497
925	14	3	44.85	-62	29	40.36	210.9369	-62.4945
926	14	4	18.23	-57	28	10.48	211.0759	-57.4696
927	14	5	36.23	-62	54	25.95	211.4009	-62.9072
928	14	5	39.2	-57	20	25.22	211.4133	-57.3403
929	14	5	43.46	-62	50	3.8	211.4311	-62.8344
930	14	5	45.32	-61	19	45.82	211.4388	-61.3294
931	14	5	49.12	-62	30	2.6	211.4547	-62.5007
932	14	5	55.9	-57	22	18.78	211.4829	-57.3719
933	14	5	56.43	-62	53	13.23	211.4851	-62.8870
934	14	6	22.76	-61	31	34.73	211.5948	-61.5263
935	14	8	39.99	-63	6	28.72	212.1666	-63.1080
936	14	10	1.76	-61	45	12.85	212.5074	-61.7536
937	14	10	35.93	-45	0	32.61	212.6497	-45.0091
938	14	12	16.38	-58	48	20.93	213.0683	-58.8058
939	14	14	21.88	-60	13	29.67	213.5912	-60.2249
940	14	15	8.91	-58	22	37.99	213.7871	-58.3772
941	14	17	4.47	-63	35	9.01	214.2686	-63.5858
942	14	17	5.36	-62	36	0.18	214.2723	-62.6000
943	14	18	0.04	-60	27	17.21	214.5002	-60.4548
944	14	19	53.9	-63	16	44.13	214.9746	-63.2789
945	14	24	23.01	-62	27	23.4	216.0959	-62.4565
946	14	26	18.77	-63	59	56.17	216.5782	-63.9989
947	14	27	20.08	-62	34	35.59	216.8336	-62.5766
948	14	27	32.3	-62	7	17	216.8846	-62.1214
949	14	27	54.13	-62	4	11.22	216.9755	-62.0698
950	14	32	24.76	-63	38	39.74	218.1031	-63.6444

951	14	35	51.28	-63	25	15.69	218.9637	-63.4210
952	14	36	5.22	-59	25	12.89	219.0218	-59.4202
953	14	37	50.17	-61	3	6.26	219.4590	-61.0517
954	14	39	32.83	-59	40	1	219.8868	-59.6669
955	14	39	35.02	-61	0	39.84	219.8959	-61.0111
956	14	41	45.8	-56	9	52.98	220.4408	-56.1647
957	14	45	2.19	-40	58	35.97	221.2591	-40.9767
958	15	1	16.53	-42	30	58.99	225.3189	-42.5164

Nicholson #956 – 15 x 15 arc minutes

CEPHEUS

#	H	M	S	D	M	S	RA DECIMAL	DEC DECIMAL
959	20	35	2.22	58	46	0.56	308.7592	58.7668
960	20	42	53.37	63	23	52.35	310.7224	63.3979
961	21	20	11.9	69	26	43.62	320.0496	69.4455
962	21	31	48.35	55	2	15.08	322.9515	55.0375
963	21	38	36.19	60	54	24.19	324.6508	60.9067
964	21	40	22.45	68	18	31.47	325.0936	68.3087
965	21	48	53.78	58	59	12.16	327.2241	58.9867
966	21	50	5.17	68	19	44.19	327.5216	68.3289
967	21	55	7.91	61	53	2.85	328.7829	61.8841
968	21	59	12.96	62	27	43.59	329.8040	62.4621
969	21	59	30.42	63	15	51.77	329.8767	63.2644
970	22	8	18.94	69	58	4.49	332.0789	69.9679
971	22	10	12.8	56	5	51.74	332.5533	56.0977
972	22	16	44.96	62	2	19.03	334.1873	62.0386
973	22	17	12.55	55	30	17.45	334.3023	55.5048
974	22	31	9.84	56	28	59.71	337.7910	56.4833
975	22	32	32.6	57	52	3.9	338.1358	57.8678
976	22	41	48.2	56	11	48.34	340.4508	56.1968
977	22	44	57.31	57	28	45.56	341.2388	57.4793
978	22	45	3.06	57	29	24.17	341.2627	57.4900
979	22	45	32.95	55	29	28.33	341.3873	55.4912
980	22	46	56.13	56	47	15.5	341.7339	56.7876
981	22	50	12.76	58	47	19.22	342.5532	58.7887
982	22	51	18.98	57	25	21.32	342.8291	57.4226
983	22	51	29.22	61	12	39.49	342.8718	61.2110
984	23	1	48.98	58	10	6.84	345.4541	58.1686
985	23	4	38.37	58	34	41.7	346.1599	58.5783
986	23	5	14.12	57	49	0.42	346.3088	57.8168
987	23	5	29.14	58	25	29.06	346.3714	58.4247
988	23	8	36.45	56	20	41.46	347.1519	56.3449

CIRCINUS

#	H	M	S	D	M	S	RA DECIMAL	DEC DECIMAL
989	13	38	30.46	-69	44	7.24	204.6269	-69.7353
990	13	38	47.05	-69	44	52.74	204.6960	-69.7480
991	13	44	22.53	-64	9	11.62	206.0939	-64.1532
992	13	47	13.92	-70	8	34.21	206.8080	-70.1428
993	13	48	25.49	-64	2	41.35	207.1062	-64.0448
994	13	48	50.76	-63	59	13.41	207.2115	-63.9871
995	13	50	8.57	-64	11	48.13	207.5357	-64.1967
996	13	54	0.37	-68	55	34.22	208.5016	-68.9262
997	13	55	57.5	-65	55	17.93	208.9896	-65.9216
998	13	57	4.38	-63	52	44.97	209.2683	-63.8792
999	13	57	15.67	-63	53	5.87	209.3153	-63.8850
1000	13	57	58.79	-68	2	59.26	209.4950	-68.0498
1001	13	58	23.3	-65	47	4.71	209.5971	-65.7846
1002	14	0	4.43	-66	12	17.54	210.0184	-66.2049
1003	14	3	18.43	-69	8	38.93	210.8268	-69.1441
1004	14	6	45.88	-64	49	40.19	211.6912	-64.8278
1005	14	10	26.21	-64	11	29.91	212.6092	-64.1916
1006	14	13	13.56	-64	10	6.57	213.3065	-64.1685
1007	14	13	42.64	-65	11	54.53	213.4276	-65.1985
1008	14	16	28.78	-71	13	31.82	214.1199	-71.2255
1009	14	18	41.46	-65	40	43.25	214.6727	-65.6787
1010	14	24	23.26	-65	9	24.73	216.0969	-65.1569
1011	14	27	18.76	-64	55	45.82	216.8282	-64.9294
1012	14	27	20.38	-64	52	18.21	216.8349	-64.8717
1013	14	27	47.28	-65	1	50.41	216.9470	-65.0307
1014	14	28	42.12	-65	52	14.19	217.1755	-65.8706
1015	14	29	38.39	-65	32	22.02	217.4100	-65.5394
1016	14	34	32.63	-66	24	9.57	218.6360	-66.4027
1017	14	40	34.44	-58	37	21.07	220.1435	-58.6225
1018	14	49	23.59	-57	26	17.14	222.3483	-57.4381
1019	14	53	2.94	-56	35	47.61	223.2623	-56.5966
1020	14	57	17.39	-55	38	16.97	224.3225	-55.6380
1021	14	58	42.42	-57	56	17.14	224.6768	-57.9381
1022	15	0	55.31	-55	3	23.15	225.2305	-55.0564
1023	15	2	30.71	-57	54	47.74	225.6280	-57.9133
1024	15	12	56.21	-55	11	48.05	228.2342	-55.1967
1025	15	18	2.6	-55	54	12.82	229.5108	-55.9036

CANIS MAJOR

#	H	M	S	D	M	S	RA DECIMAL	DEC DECIMAL
1026	6	9	16.88	-17	17	29.74	92.3203	-17.2916
1027	6	18	0.85	-29	41	28.17	94.5035	-29.6912
1028	6	26	25.17	-15	35	12.66	96.6049	-15.5869
1029	6	33	10.6	-25	29	51.53	98.2942	-25.4976
1030	6	35	35.86	-19	22	27.89	98.8994	-19.3744
1031	6	39	33.16	-31	48	37.19	99.8882	-31.8103
1032	6	41	47.17	-19	40	25.11	100.4466	-19.6736
1033	6	45	33	-24	16	0.69	101.3875	-24.2669
1034	6	47	12.57	-12	3	25.51	101.8024	-12.0571
1035	6	47	56.77	-22	8	32.09	101.9865	-22.1422
1036	6	50	43.21	-20	56	15.12	102.6801	-20.9375
1037	6	52	57.89	-27	5	10.11	103.2412	-27.0861
1038	6	53	1.98	-31	27	22.77	103.2582	-31.4563
1039	6	54	30.6	-23	6	22.57	103.6275	-23.1063
1040	6	56	37.86	-28	5	42.37	104.1578	-28.0951
1041	6	58	28.48	-20	5	51.32	104.6187	-20.0976
1042	6	58	30.92	-21	2	19.17	104.6288	-21.0387
1043	6	59	3.11	-26	13	55.4	104.7630	-26.2321
1044	6	59	29.44	-23	29	28.5	104.8727	-23.4912
1045	7	0	51.8	-22	37	3.2	105.2158	-22.6176
1046	7	2	52.64	-17	29	40.14	105.7193	-17.4945
1047	7	3	31.4	-29	51	21.21	105.8808	-29.8559
1048	7	3	53.35	-24	7	6.92	105.9723	-24.1186
1049	7	4	9.89	-17	22	28.99	106.0412	-17.3747
1050	7	5	18.53	-25	44	3.99	106.3272	-25.7344
1051	7	5	23.16	-29	9	40.46	106.3465	-29.1612
1052	7	5	54.26	-26	13	51.2	106.4761	-26.2309
1053	7	6	4.65	-29	32	26.65	106.5194	-29.5407
1054	7	7	58.46	-15	25	6.39	106.9936	-15.4184
1055	7	8	24.91	-17	51	1.81	107.1038	-17.8505
1056	7	8	41.5	-24	2	40.29	107.1729	-24.0445
1057	7	9	24.75	-16	16	37.31	107.3531	-16.2770
1058	7	9	32.48	-15	12	12.55	107.3853	-15.2035
1059	7	10	7.88	-24	35	50.12	107.5328	-24.5973
1060	7	10	23.94	-19	44	0.4	107.5998	-19.7334
1061	7	10	35.89	-16	12	45.47	107.6495	-16.2126
1062	7	10	41.84	-16	10	54.23	107.6743	-16.1817
1063	7	10	42.72	-29	4	31.34	107.6780	-29.0754
1064	7	11	24.68	-20	15	48.98	107.8528	-20.2636

1065	7	11	27.55	-24	42	54.46	107.8648	-24.7151
1066	7	14	20.03	-19	0	55.47	108.5835	-19.0154
1067	7	15	12.03	-21	55	24.42	108.8001	-21.9235
1068	7	15	25.38	-18	22	22.54	108.8557	-18.3729
1069	7	16	29.22	-22	20	37	109.1217	-22.3436
1070	7	16	30.88	-14	52	28.95	109.1287	-14.8747
1071	7	17	1.38	-22	42	42.07	109.2557	-22.7117
1072	7	17	13.33	-31	30	20.51	109.3056	-31.5057
1073	7	17	39.28	-23	18	22.25	109.4137	-23.3062
1074	7	19	31.09	-15	7	48.4	109.8795	-15.1301
1075	7	19	51.41	-24	4	0.74	109.9642	-24.0669
1076	7	20	1.45	-28	32	33.53	110.0060	-28.5426
1077	7	20	38.68	-14	52	19.83	110.1612	-14.8722
1078	7	20	56.15	-30	16	6.22	110.2340	-30.2684
1079	7	21	4.35	-30	51	53.58	110.2681	-30.8649
1080	7	21	9.36	-14	39	25.87	110.2890	-14.6572
1081	7	21	11.96	-30	43	33.75	110.2999	-30.7260
1082	7	21	15.54	-14	43	54.06	110.3148	-14.7317
1083	7	21	22.85	-30	42	9.95	110.3452	-30.7028
1084	7	21	23.06	-14	46	1.12	110.3461	-14.7670
1085	7	21	43.42	-15	12	7.3	110.4309	-15.2020
1086	7	22	32.15	-23	42	43.22	110.6339	-23.7120
1087	7	22	38.22	-21	47	14.36	110.6593	-21.7873
1088	7	22	41.65	-23	44	39.47	110.6735	-23.7443
1089	7	22	53.18	-16	43	3.29	110.7216	-16.7176
1090	7	23	28.26	-18	5	15.03	110.8678	-18.0875
1091	7	24	38.37	-16	24	10.15	111.1599	-16.4028
1092	7	24	40.95	-20	23	1.98	111.1706	-20.3839
1093	7	24	44.6	-15	53	17.51	111.1858	-15.8882
1094	7	25	34.56	-14	6	8.11	111.3940	-14.1023
1095	7	26	21.07	-13	8	21.69	111.5878	-13.1394
1096	7	27	53.33	-20	8	56.22	111.9722	-20.1490
1097	7	27	59.17	-19	42	21.17	111.9965	-19.7059
1098	7	30	56.37	-24	40	1.8	112.7349	-24.6672
1099	7	32	10.71	-26	53	36.22	113.0446	-26.8934
1100	7	34	8.11	-28	42	21.98	113.5338	-28.7061

CANIS MINOR

#	H	M	S	D	M	S	RA DECIMAL	DEC DECIMAL
1101	7	6	11.78	12	48	32.4	106.5491	12.8090
1102	7	6	54.14	11	26	55.88	106.7256	11.4489
1103	7	13	51.22	8	9	1.74	108.4634	8.1505
1104	7	15	28.71	12	12	53.12	108.8696	12.2148
1105	7	16	13.72	10	4	29.06	109.0572	10.0747
1106	7	17	15.9	6	39	9.8	109.3162	6.6527
1107	7	18	16.62	8	27	3.88	109.5693	8.4511
1108	7	19	59.38	0	15	20.27	109.9974	0.2556
1109	7	20	17.82	1	4	35.41	110.0742	1.0765
1110	7	20	36.68	9	42	44.66	110.1529	9.7124
1111	7	20	56.23	3	26	4.84	110.2343	3.4347
1112	7	22	46.53	1	10	57.46	110.6939	1.1826
1113	7	26	59.57	-3	21	30.47	111.7482	-3.3585
1114	7	28	35.67	-1	17	33.52	112.1486	-1.2926
1115	7	32	12.89	3	37	22.33	113.0537	3.6229
1116	7	32	46.4	-2	45	42.21	113.1933	-2.7617
1117	7	35	32.12	0	33	38.07	113.8838	0.5606
1118	7	36	22.47	0	5	30.45	114.0936	-0.0918
1119	7	42	8.48	-3	47	28.43	115.5353	-3.7912
1120	7	47	4.25	1	58	25.07	116.7677	1.9736

Nicholson #1120 – 15 x 15 arc minutes

COLUMBA

#	H	M	S	D	M	S	RA DECIMAL	DEC DECIMAL
1121	5	50	32.98	-29	11	54.91	87.6374	-29.1986
1122	6	25	42.8	-36	20	38.25	96.4283	-36.3440

CORONA AUSTRALIS

#	H	M	S	D	M	S	RA DECIMAL	DEC DECIMAL
1123	17	59	0.17	-40	58	27.76	269.7507	-40.9744
1124	18	3	6.85	-43	21	39.94	270.7785	-43.3611
1125	18	6	29.87	-41	41	7.83	271.6245	-41.6855
1126	18	10	49.78	-40	31	51.71	272.7074	-40.5310
1127	18	11	2.94	-43	10	46.13	272.7622	-43.1795
1128	18	12	0.19	-42	37	4.5	273.0008	-42.6179
1129	18	12	1.99	-40	33	19.14	273.0083	-40.5553
1130	18	12	2.59	-42	41	52.17	273.0108	-42.6978
1131	18	12	7.66	-42	40	10.04	273.0319	-42.6695
1132	18	12	23.27	-40	23	5.32	273.0970	-40.3848
1133	18	21	24.96	-42	56	0.45	275.3540	-42.9335
1134	18	22	50.44	-40	44	18.85	275.7102	-40.7386
1135	18	24	12.98	-42	12	20.14	276.0541	-42.2056
1136	18	24	58.39	-42	1	19.99	276.2433	-42.0222
1137	18	25	5.92	-41	58	20.37	276.2747	-41.9723
1138	18	26	12.66	-41	12	28.88	276.5527	-41.2080
1139	18	28	14.5	-41	1	4.98	277.0604	-41.0180
1140	18	30	22.65	-41	59	34.65	277.5944	-41.9930
1141	18	40	2.03	-42	18	40.87	280.0085	-42.3114
1142	18	43	38.37	-45	49	26.39	280.9099	-45.8240
1143	18	45	43.76	-40	7	52.48	281.4324	-40.1312
1144	18	46	26.16	-42	1	54.51	281.6090	-42.0318
1145	18	48	41.9	-41	42	24.64	282.1746	-41.7068
1146	18	53	38.23	-40	57	36.73	283.4093	-40.9602
1147	18	56	21.48	-39	54	21.38	284.0895	-39.9059
1148	19	11	41.71	-36	23	34.59	287.9238	-36.3929

CRUX

#	H	M	S	D	M	S	RA DECIMAL	DEC DECIMAL
1149	11	50	17.45	-59	50	2.82	177.5727	-59.8341
1150	11	51	0.6	-60	4	0.6	177.7525	-60.0668
1151	11	51	5.96	-63	28	57.32	177.7748	-63.4826
1152	11	51	14.35	-60	10	16.99	177.8098	-60.1714
1153	11	51	19.28	-63	28	22.91	177.8303	-63.4730
1154	11	51	35.08	-60	9	16.42	177.8962	-60.1546
1155	11	52	1.53	-63	9	10.01	178.0064	-63.1528
1156	11	52	48.66	-60	35	45.95	178.2027	-60.5961
1157	11	53	14.2	-60	1	7.87	178.3092	-60.0189
1158	11	54	13.62	-63	33	46.6	178.5567	-63.5629
1159	11	54	33.02	-59	58	20.44	178.6376	-59.9723
1160	11	54	35.67	-63	42	10.95	178.6486	-63.7030
1161	11	54	36.37	-60	49	18.7	178.6515	-60.8219
1162	11	54	37	-63	36	40.32	178.6542	-63.6112
1163	11	54	46.04	-60	39	37.26	178.6918	-60.6603
1164	11	54	47.45	-61	51	2.83	178.6977	-61.8508
1165	11	54	49.18	-60	39	6.57	178.7049	-60.6518
1166	11	54	52.75	-61	52	49.77	178.7198	-61.8805
1167	11	54	55.38	-63	43	39.98	178.7307	-63.7278
1168	11	55	1.01	-61	50	3.5	178.7542	-61.8343
1169	11	55	46.07	-63	10	40.75	178.9420	-63.1780
1170	11	55	58.01	-63	11	41.96	178.9917	-63.1950
1171	11	56	7.01	-61	0	2.95	179.0292	-61.0008
1172	11	56	7.33	-63	14	9.7	179.0305	-63.2360
1173	11	56	14.81	-60	58	45.24	179.0617	-60.9792
1174	11	56	18.98	-62	3	41.25	179.0791	-62.0615
1175	11	56	26.83	-60	42	38.95	179.1118	-60.7108
1176	11	56	32.82	-62	4	44.67	179.1368	-62.0791
1177	11	56	54.76	-63	49	55.16	179.2282	-63.8320
1178	11	58	10.2	-59	32	59.72	179.5425	-59.5499
1179	11	58	24.37	-59	29	3.62	179.6016	-59.4843
1180	11	58	36.5	-59	29	24.92	179.6521	-59.4903
1181	11	58	55.36	-59	29	24.95	179.7307	-59.4903
1182	11	59	13.12	-59	17	58.06	179.8047	-59.2995
1183	11	59	23.44	-62	7	7.6	179.8477	-62.1188
1184	12	0	22.08	-62	3	18.56	180.0920	-62.0552
1185	12	0	39.28	-62	0	21.63	180.1637	-62.0060
1186	12	0	41.3	-61	28	31.73	180.1721	-61.4755
1187	12	0	43.69	-62	0	28.52	180.1820	-62.0079

1188	12	0	47.38	-60	27	53.28	180.1974	-60.4648
1189	12	0	53.04	-61	21	52.95	180.2210	-61.3647
1190	12	1	1.53	-60	24	49.84	180.2564	-60.4138
1191	12	1	17.9	-61	45	12.89	180.3246	-61.7536
1192	12	1	48.73	-59	56	6.61	180.4531	-59.9352
1193	12	2	47.95	-59	27	6.84	180.6998	-59.4519
1194	12	4	35.62	-64	30	23.05	181.1484	-64.5064
1195	12	4	39.34	-64	33	17.67	181.1639	-64.5549
1196	12	5	38.77	-61	56	52.39	181.4116	-61.9479
1197	12	6	13.32	-64	32	6.89	181.5555	-64.5352
1198	12	7	36.83	-64	5	40.07	181.9035	-64.0945
1199	12	8	42.02	-64	31	56.4	182.1751	-64.5323
1200	12	9	42.76	-62	11	57.94	182.4282	-62.1994
1201	12	9	52.29	-63	49	48.6	182.4679	-63.8302
1202	12	11	10.07	-58	21	22.99	182.7919	-58.3564
1203	12	11	33.61	-57	39	13.97	182.8901	-57.6539
1204	12	14	5.22	-61	18	31.11	183.5217	-61.3086
1205	12	14	16.46	-61	20	38.82	183.5686	-61.3441
1206	12	14	26.76	-58	51	0.21	183.6115	-58.8501
1207	12	15	47.57	-58	24	55.33	183.9482	-58.4154
1208	12	16	20.04	-61	3	9.17	184.0835	-61.0525
1209	12	17	29.96	-63	9	14.75	184.3748	-63.1541
1210	12	18	29.73	-60	54	52.35	184.6239	60.9145
1211	12	26	22.31	-58	37	53.89	186.5930	-58.6316
1212	12	26	48.65	-62	53	44.06	186.7027	-62.8956
1213	12	27	25.29	-62	41	46.1	186.8554	-62.6961
1214	12	28	25.63	-59	9	30.36	187.1068	-59.1584
1215	12	28	54.89	-62	59	1.6	187.2287	-62.9838
1216	12	28	59.59	-62	6	34.88	187.2483	-62.1097
1217	12	35	36.93	-57	37	21.88	188.9039	-57.6227
1218	12	35	54.06	-59	32	2.21	188.9752	-59.5339
1219	12	46	56.62	-59	20	35.99	191.7359	-59.3433
1220	12	48	3.29	-59	26	16.26	192.0137	-59.4379
1221	12	48	4.39	-60	2	11.57	192.0183	-60.0365
1222	12	48	9.27	-59	23	48.94	192.0386	-59.3969
1223	12	53	17.98	-59	37	16	193.3249	-59.6211
1224	12	58	19.9	-60	39	7.68	194.5829	-60.6521
1225	13	2	7.84	-60	47	7.02	195.5327	-60.7853

CYGNUS

#	H	M	S	D	M	S	RA DECIMAL	DEC DECIMAL
1226	19	20	3.36	35	32	24.91	290.0140	35.5403
1227	19	26	44.23	39	57	41.41	291.6843	39.9615
1228	19	27	44.2	34	36	40.6	291.9342	34.6113
1229	19	27	52.02	35	29	34.06	291.9668	35.4928
1230	19	28	2.54	34	33	26.68	292.0106	34.5574
1231	19	28	29.67	35	17	4.9	292.1236	35.2847
1232	19	28	35.68	41	1	5.89	292.1487	41.0183
1233	19	29	38.01	40	33	10.04	292.4084	40.5528
1234	19	31	8.89	40	31	5.52	292.7870	40.5182
1235	19	31	39.47	35	9	51.19	292.9144	35.1642
1236	19	32	1.79	38	44	50.49	293.0075	38.7474
1237	19	32	7.63	34	44	47.87	293.0318	34.7466
1238	19	32	22.45	34	17	2.54	293.0935	34.2840
1239	19	32	39.35	54	18	26.19	293.1639	54.3073
1240	19	33	9.29	39	43	53.69	293.2887	39.7316
1241	19	33	21.41	39	23	27.3	293.3392	39.3909
1242	19	33	24.49	34	16	32.26	293.3520	34.2756
1243	19	33	34.04	36	15	21.28	293.3918	36.2559
1244	19	33	42.76	31	54	48.55	293.4282	31.9135
1245	19	33	43.88	36	14	35.14	293.4328	36.2431
1246	19	34	10.12	32	51	21.98	293.5422	32.8561
1247	19	34	50.27	31	54	51.8	293.7094	31.9144
1248	19	34	55.88	41	54	2.98	293.7328	41.9008
1249	19	35	1.06	33	29	56.07	293.7544	33.4989
1250	19	36	6.23	31	23	58.47	294.0260	31.3996
1251	19	36	18.11	31	30	27.8	294.0755	31.5077
1252	19	36	22.43	31	26	19.88	294.0935	31.4389
1253	19	36	24.25	37	4	50.42	294.1011	37.0807
1254	19	36	24.69	31	12	12.76	294.1029	31.2035
1255	19	36	28.16	37	1	29.25	294.1174	37.0248
1256	19	36	33.25	31	29	50.97	294.1385	31.4975
1257	19	36	39.29	43	27	35.41	294.1637	43.4598
1258	19	36	42.93	31	17	11.56	294.1789	31.2865
1259	19	36	52.05	31	20	3.01	294.2169	31.3342
1260	19	37	12.59	37	11	52.1	294.3025	37.1978
1261	19	37	17.79	31	28	33.15	294.3241	31.4759
1262	19	37	18.54	34	2	18.57	294.3273	34.0385
1263	19	37	33.45	33	31	29.09	294.3894	33.5247
1264	19	37	43.48	40	35	38.38	294.4312	40.5940

1265	19	37	47.07	32	27	35.89	294.4461	32.4600
1266	19	37	47.23	33	30	7.73	294.4468	33.5021
1267	19	37	51.85	39	49	46.08	294.4660	39.8295
1268	19	38	9.34	34	12	12.11	294.5389	34.2034
1269	19	38	9.98	32	24	9.21	294.5416	32.4026
1270	19	38	10.33	35	59	53.88	294.5430	35.9983
1271	19	38	23.19	35	48	49.29	294.5966	35.8137
1272	19	38	28.1	54	18	32.38	294.6171	54.3090
1273	19	38	29.6	35	54	44.67	294.6233	35.9124
1274	19	38	58.04	42	2	22.89	294.7418	42.0397
1275	19	39	2.86	35	0	5.71	294.7619	35.0016
1276	19	39	5.33	32	21	54.28	294.7722	32.3651
1277	19	39	6.06	38	48	46.46	294.7752	38.8129
1278	19	39	9.41	32	45	47.66	294.7892	32.7632
1279	19	39	17.28	32	44	51.17	294.8220	32.7475
1280	19	39	20.75	38	47	10.53	294.8365	38.7863
1281	19	39	42.93	32	58	7.88	294.9289	32.9689
1282	19	39	46.69	30	17	32.99	294.9445	30.2925
1283	19	39	54.22	32	11	17.13	294.9759	32.1881
1284	19	40	4.63	30	44	8.53	295.0193	30.7357
1285	19	40	17.4	30	59	15.64	295.0725	30.9877
1286	19	40	33.39	35	17	49.55	295.1391	35.2971
1287	19	41	6.95	33	54	19.56	295.2790	33.9054
1288	19	41	7.66	34	30	3.52	295.2819	34.5010
1289	19	41	8.55	32	6	38.26	295.2856	32.1106
1290	19	41	10.91	33	50	54.67	295.2955	33.8485
1291	19	41	13.05	32	4	6.95	295.3044	32.0686
1292	19	41	51.6	31	2	22.18	295.4650	31.0395
1293	19	41	58.36	35	43	3.97	295.4931	35.7178
1294	19	42	0.85	31	0	29.98	295.5036	31.0083
1295	19	42	5.39	49	21	23.05	295.5225	49.3564
1296	19	42	8.75	41	44	28.91	295.5364	41.7414
1297	19	42	16.04	31	30	4.3	295.5668	31.5012
1298	19	42	29.31	45	9	53.08	295.6221	45.1647
1299	19	42	29.77	31	12	55.34	295.6240	31.2154
1300	19	43	22.9	36	16	36.22	295.8454	36.2767
1301	19	43	27.66	35	47	21.8	295.8653	35.7894
1302	19	43	35.97	35	51	24.01	295.8999	35.8567
1303	19	43	39.41	35	52	26.91	295.9142	35.8741
1304	19	43	44.98	32	0	0.7	295.9374	32.0002
1305	19	43	46.4	35	3	29.15	295.9433	35.0581
1306	19	43	48.46	30	39	13.93	295.9519	30.6539
1307	19	44	31.93	30	52	11.94	296.1330	30.8700

1308	19	44	36.42	30	10	0.4	296.1518	30.1668
1309	19	45	22.51	36	55	49.37	296.3438	36.9304
1310	19	45	32.95	31	26	49.29	296.3873	31.4470
1311	19	45	39.06	30	5	11.02	296.4128	30.0864
1312	19	45	40.81	31	26	7.6	296.4200	31.4354
1313	19	45	54.44	32	20	10.26	296.4768	32.3362
1314	19	46	16.2	32	21	11.3	296.5675	32.3531
1315	19	46	16.26	40	59	52.48	296.5677	40.9979
1316	19	46	18.17	39	52	12.1	296.5757	39.8700
1317	19	46	23.52	33	27	25.12	296.5980	33.4570
1318	19	46	30.32	48	27	13.14	296.6263	48.4537
1319	19	46	35.39	33	28	20.14	296.6475	33.4723
1320	19	46	41.06	32	39	45.92	296.6711	32.6628
1321	19	46	42.23	33	25	39.18	296.6760	33.4276
1322	19	46	43.28	30	2	28.18	296.6803	30.0412
1323	19	46	43.4	33	28	44.36	296.6808	33.4790
1324	19	46	43.96	32	38	44.32	296.6832	32.6456
1325	19	46	44.11	30	3	6.02	296.6838	30.0517
1326	19	46	46.7	32	42	56.72	296.6946	32.7158
1327	19	46	46.86	30	5	59.06	296.6953	30.0997
1328	19	46	53.32	32	38	31.52	296.7221	32.6421
1329	19	46	58.99	38	35	12.88	296.7458	38.5869
1330	19	46	59.1	33	28	52.32	296.7463	33.4812
1331	19	47	6.28	28	48	12.59	296.7761	28.8035
1332	19	47	23.44	37	48	2.06	296.8477	37.8006
1333	19	47	27.06	31	0	37.68	296.8627	31.0105
1334	19	47	49.89	28	53	5.35	296.9579	28.8848
1335	19	47	50.88	34	58	25.42	296.9620	34.9737
1336	19	47	55.11	39	58	24.05	296.9796	39.9733
1337	19	48	9.21	29	55	24.96	297.0384	29.9236
1338	19	48	9.43	32	7	26.96	297.0393	32.1242
1339	19	48	10.3	29	53	28.1	297.0429	29.8911
1340	19	48	16.71	40	31	30.52	297.0696	40.5251
1341	19	48	16.79	35	28	16.27	297.0699	35.4712
1342	19	48	18.18	40	30	54.79	297.0757	40.5152
1343	19	48	22.77	43	4	40.19	297.0949	43.0778
1344	19	48	24.91	44	56	56.26	297.1038	44.9490
1345	19	48	30.59	47	48	9.83	297.1275	47.8027
1346	19	48	35.39	35	59	50.66	297.1474	35.9974
1347	19	48	36.51	43	6	32.25	297.1521	43.1090
1348	19	48	43.62	44	55	33.02	297.1817	44.9258
1349	19	49	11.82	35	52	4.36	297.2993	35.8679
1350	19	49	37.62	38	53	42.91	297.4067	38.8953

1351	19	49	43.17	38	54	44.7	297.4299	38.9124
1352	19	49	43.96	28	33	4.87	297.4332	28.5514
1353	19	49	48.14	35	17	13.28	297.4506	35.2870
1354	19	50	1.2	28	13	21.53	297.5050	28.2226
1355	19	50	30.58	47	32	7.9	297.6274	47.5355
1356	19	50	31.4	38	37	44.11	297.6309	38.6289
1357	19	50	46.18	30	56	55.6	297.6924	30.9488
1358	19	50	50.67	34	26	54.74	297.7111	34.4485
1359	19	51	18.31	37	24	48.77	297.8263	37.4135
1360	19	51	27.04	28	54	52.5	297.8627	28.9146
1361	19	51	32.17	35	57	37.06	297.8840	35.9603
1362	19	51	49.11	36	33	42.1	297.9546	36.5617
1363	19	51	56.39	31	10	22.86	297.9849	31.1730
1364	19	52	10.12	28	3	38.69	298.0422	28.0607
1365	19	52	14.76	36	9	33.56	298.0615	36.1593
1366	19	53	53.95	44	17	36.81	298.4748	44.2936
1367	19	54	11.56	32	54	45.76	298.5482	32.9127
1368	19	54	15.63	32	52	25.57	298.5651	32.8738
1369	19	54	24.56	38	22	26.5	298.6023	38.3740
1370	19	54	36.73	37	24	9.08	298.6531	37.4025
1371	19	54	39.47	32	56	2.83	298.6645	32.9341
1372	19	55	22.51	37	40	37.28	298.8438	37.6770
1373	19	55	28.67	40	52	56.94	298.8695	40.8825
1374	19	55	29.89	37	37	16.54	298.8745	37.6213
1375	19	55	42.82	38	24	10.81	298.9284	38.4030
1376	19	55	47.01	43	6	5.51	298.9459	43.1015
1377	19	55	49.64	41	36	17.07	298.9568	41.6047
1378	19	55	58.62	35	49	28.61	298.9943	35.8246
1379	19	56	8.2	57	17	30.62	299.0342	57.2918
1380	19	56	9.31	38	51	39.94	299.0388	38.8611
1381	19	56	20.23	39	34	45.94	299.0843	39.5794
1382	19	56	20.28	44	53	46.73	299.0845	44.8963
1383	19	56	35.4	35	47	45.88	299.1475	35.7961
1384	19	56	41.83	42	20	40.51	299.1743	42.3446
1385	19	56	57.63	41	3	17.37	299.2401	41.0548
1386	19	57	16.29	35	59	24.37	299.3179	35.9901
1387	19	57	45.14	33	15	32.84	299.4381	33.2591
1388	19	57	56.87	37	53	33.62	299.4870	37.8927
1389	19	58	15.34	39	8	34.19	299.5639	39.1428
1390	19	58	25.53	38	33	50.98	299.6064	38.5642
1391	19	58	30.76	40	21	14.79	299.6282	40.3541
1392	19	58	35.36	38	35	57.82	299.6474	38.5994
1393	19	58	39.23	41	47	49.41	299.6635	41.7971

1394	19	58	45.81	35	37	17.56	299.6909	35.6215
1395	19	58	51.75	37	44	13.02	299.7156	37.7370
1396	19	59	13.59	35	29	49.85	299.8066	35.4972
1397	19	59	30.3	38	55	51.05	299.8763	38.9308
1398	19	59	43.47	37	13	11.25	299.9311	37.2198
1399	19	59	47.53	44	32	20.67	299.9480	44.5391
1400	19	59	49.32	38	56	5.1	299.9555	38.9347
1401	19	59	52.03	33	9	15.72	299.9668	33.1544
1402	19	59	54.46	38	3	58.92	299.9769	38.0664
1403	19	59	59.47	33	10	44	299.9978	33.1789
1404	20	0	3.43	38	8	53.78	300.0143	38.1483
1405	20	0	6.6	40	7	50.58	300.0275	40.1307
1406	20	0	7.88	39	33	49.61	300.0328	39.5638
1407	20	0	8.88	36	18	24.03	300.0370	36.3067
1408	20	0	15.32	31	51	21.7	300.0638	31.8560
1409	20	0	15.93	40	6	37.05	300.0664	40.1103
1410	20	0	27.06	32	47	57.13	300.1128	32.7992
1411	20	0	30.44	32	42	48.31	300.1268	32.7134
1412	20	0	42.59	36	34	16.88	300.1775	36.5714
1413	20	0	51.75	41	15	16.58	300.2156	41.2546
1414	20	0	54.48	33	0	44.98	300.2270	33.0125
1415	20	1	9.24	41	12	20.44	300.2885	41.2057
1416	20	1	42.02	58	10	10.79	300.4251	58.1697
1417	20	2	3.14	43	7	37.04	300.5131	43.1270
1418	20	2	50.62	47	4	6.58	300.7109	47.0685
1419	20	2	56.3	43	15	33.39	300.7346	43.2593
1420	20	3	6.39	40	9	44.97	300.7766	40.1625
1421	20	3	26.54	58	10	9.91	300.8606	58.1694
1422	20	3	41.09	50	52	19.53	300.9212	50.8721
1423	20	4	32.77	34	59	32.09	301.1366	34.9922
1424	20	4	36.45	38	50	23.81	301.1519	38.8399
1425	20	5	30.08	36	39	8.31	301.3753	36.6523
1426	20	5	36.49	47	31	49.64	301.4021	47.5305
1427	20	5	37.91	50	30	7.52	301.4080	50.5021
1428	20	5	37.98	52	25	54.71	301.4083	52.4319
1429	20	5	38.09	50	30	5.09	301.4087	50.5014
1430	20	5	45.85	47	33	3.38	301.4410	47.5509
1431	20	5	53.25	50	31	34.05	301.4719	50.5261
1432	20	6	57.75	34	37	54.7	301.7406	34.6319
1433	20	7	5.96	49	57	53.57	301.7748	49.9649
1434	20	7	28.04	49	43	5.14	301.8668	49.7181
1435	20	7	38.85	38	52	12.38	301.9119	38.8701
1436	20	7	46.99	37	1	3.26	301.9458	37.0176

1437	20	9	13.23	43	5	22.52	302.3051	43.0896
1438	20	9	45.06	39	33	21.36	302.4377	39.5559
1439	20	9	59.67	50	37	7.8	302.4986	50.6188
1440	20	10	5.12	51	21	30.36	302.5213	51.3584
1441	20	10	14.36	50	36	23.76	302.5598	50.6066
1442	20	11	8.64	38	57	43.84	302.7860	38.9622
1443	20	12	16.5	33	34	4.77	303.0688	33.5680
1444	20	13	20.38	40	43	22.75	303.3349	40.7230
1445	20	13	25.45	35	6	14.88	303.3561	35.1041
1446	20	14	14.16	30	48	40.63	303.5590	30.8113
1447	20	14	15.13	44	16	53.08	303.5630	44.2814
1448	20	14	16.04	30	51	6.34	303.5668	30.8518
1449	20	14	20.67	42	35	13.83	303.5861	42.5872
1450	20	14	25.3	46	46	59.96	303.6054	46.7833
1451	20	16	11.92	55	7	17.34	304.0497	55.1215
1452	20	16	27.12	35	41	57.29	304.1130	35.6992
1453	20	17	7.54	41	57	26.5	304.2814	41.9574
1454	20	17	34.56	49	24	17.02	304.3940	49.4047
1455	20	18	36.37	39	55	32.21	304.6515	39.9256
1456	20	19	31.13	32	2	49.98	304.8797	32.0472
1457	20	19	36.84	33	47	34.94	304.9035	33.7930
1458	20	19	48.74	56	28	36.31	304.9531	56.4768
1459	20	19	55.58	33	36	31.63	304.9816	33.6088
1460	20	20	11.12	51	32	29.28	305.0463	51.5415
1461	20	23	7.08	53	20	19.59	305.7795	53.3388
1462	20	23	33.77	47	29	0.17	305.8907	47.4834
1463	20	23	34.73	34	49	6.67	305.8947	34.8185
1464	20	23	38.85	50	20	52.31	305.9119	50.3479
1465	20	23	53.13	44	37	24.13	305.9714	44.6234
1466	20	23	56.62	52	31	32.86	305.9859	52.5258
1467	20	23	57.15	53	21	3.39	305.9881	53.3509
1468	20	24	24.26	51	42	39.82	306.1011	51.7111
1469	20	24	24.26	29	25	49	306.1011	29.4303
1470	20	25	7.78	32	3	23.24	306.2824	32.0565
1471	20	25	14.31	35	21	32.21	306.3096	35.3589
1472	20	25	49.06	40	51	34.83	306.4544	40.8597
1473	20	25	49.07	53	3	38.92	306.4545	53.0608
1474	20	26	7.94	47	59	21.87	306.5331	47.9894
1475	20	26	38.09	53	46	51.66	306.6587	53.7810
1476	20	27	18.77	53	55	3.29	306.8282	53.9176
1477	20	27	40.92	52	30	11.44	306.9205	52.5032
1478	20	27	48.41	40	36	28.82	306.9517	40.6080
1479	20	28	11.15	45	3	4.03	307.0464	45.0511

1480	20	29	12.16	54	18	48.03	307.3007	54.3133
1481	20	29	22.2	54	21	19.17	307.3425	54.3553
1482	20	29	27.15	47	3	57.37	307.3631	47.0659
1483	20	29	41.3	46	2	40.35	307.4221	46.0445
1484	20	31	5.15	51	2	5.25	307.7714	51.0348
1485	20	31	14.32	49	0	6.86	307.8097	49.0019
1486	20	31	23.08	32	52	55.5	307.8462	32.8821
1487	20	31	43.19	45	12	19.31	307.9300	45.2054
1488	20	31	50.04	54	11	54.53	307.9585	54.1985
1489	20	31	53.34	35	18	19.94	307.9722	35.3055
1490	20	31	56.3	45	11	0.68	307.9846	45.1835
1491	20	31	59.62	35	16	4.66	307.9984	35.2680
1492	20	32	0.98	45	13	46.03	308.0041	45.2295
1493	20	32	30.84	32	32	11.38	308.1285	32.5365
1494	20	32	38.05	51	17	44.67	308.1585	51.2957
1495	20	32	58.13	39	14	31.05	308.2422	39.2420
1496	20	33	54.58	51	23	26.68	308.4774	51.3907
1497	20	34	50.92	32	27	33.39	308.7122	32.4593
1498	20	34	56.71	55	24	46.01	308.7363	55.4128
1499	20	35	0.73	52	23	21.5	308.7530	52.3893
1500	20	38	3.05	47	49	26.12	309.5127	47.8239
1501	20	38	26.25	31	18	36.98	309.6094	31.3103
1502	20	38	53.48	46	32	20.58	309.7228	46.5391
1503	20	39	47.11	54	47	13.84	309.9463	54.7872
1504	20	40	27.04	47	18	21.62	310.1127	47.3060
1505	20	40	52.13	47	41	40.35	310.2172	47.6945
1506	20	42	15.95	34	30	10.59	310.5664	34.5029
1507	20	42	24.92	45	48	44.24	310.6038	45.8123
1508	20	42	33.03	47	41	28.78	310.6376	47.6913
1509	20	42	59.38	31	43	58.11	310.7474	31.7328
1510	20	43	12.93	34	6	1.1	310.8039	34.1003
1511	20	43	43.84	47	57	2.1	310.9327	47.9506
1512	20	46	0.46	47	24	13.68	311.5019	47.4038
1513	20	46	34.19	34	52	1.02	311.6425	34.8670
1514	20	46	45.96	36	25	31.97	311.6915	36.4255
1515	20	46	56.35	32	51	50.69	311.7348	32.8641
1516	20	48	44.07	33	40	12.34	312.1836	33.6701
1517	20	49	15.07	33	59	8.73	312.3128	33.9858
1518	20	49	23.53	39	20	44.54	312.3480	39.3457
1519	20	49	48.71	35	15	46.49	312.4530	35.2629
1520	20	49	52.21	33	39	23.99	312.4675	33.6567
1521	20	49	54.64	28	21	22.93	312.4777	28.3564
1522	20	50	5.74	34	15	38.4	312.5239	34.2607

1523	20	50	51.8	33	54	18.17	312.7158	33.9050
1524	20	51	11.49	48	57	42.86	312.7979	48.9619
1525	20	51	12.42	34	46	36.34	312.8018	34.7768
1526	20	52	11.36	35	12	48.85	313.0473	35.2136
1527	20	52	14.29	29	33	1.69	313.0596	29.5505
1528	20	52	24.27	35	11	49.46	313.1011	35.1971
1529	20	52	30	34	42	35.15	313.1250	34.7098
1530	20	52	45	46	44	46.94	313.1875	46.7464
1531	20	52	45.28	33	14	31.47	313.1887	33.2421
1532	20	52	50.01	46	46	21.8	313.2084	46.7727
1533	20	53	31.97	36	40	7.19	313.3832	36.6687
1534	20	54	0.24	34	58	39.29	313.5010	34.9776
1535	20	54	10.6	37	3	38.3	313.5442	37.0606
1536	20	54	13.75	31	21	16.2	313.5573	31.3545
1537	20	54	21.07	37	3	55.15	313.5878	37.0653
1538	20	55	33.78	36	39	25.07	313.8908	36.6570
1539	20	56	33.49	34	1	7.39	314.1395	34.0187
1540	20	56	45.47	30	12	29.13	314.1895	30.2081
1541	20	57	47.22	36	39	27.74	314.4467	36.6577
1542	20	58	30.06	45	1	37.98	314.6252	45.0272
1543	20	58	36.43	45	56	21.76	314.6518	45.9394
1544	20	58	36.57	44	59	36.14	314.6524	44.9934
1545	20	58	40.31	45	56	8.44	314.6680	45.9357
1546	20	59	22.42	35	28	28.57	314.8434	35.4746
1547	20	59	48.36	37	3	40.55	314.9515	37.0613
1548	21	0	58.67	46	22	13	315.2445	46.3703
1549	21	1	0.23	45	28	7.83	315.2509	45.4688
1550	21	1	46.04	32	59	45.1	315.4418	32.9959
1551	21	2	57.23	35	32	38.88	315.7384	35.5441
1552	21	3	17.16	33	41	54.1	315.8215	33.6984
1553	21	4	39.1	35	31	58.17	316.1629	35.5328
1554	21	4	46.16	38	44	57.79	316.1923	38.7494
1555	21	4	49.19	35	31	53.5	316.2050	35.5315
1556	21	4	49.98	50	1	33.64	316.2083	50.0260
1557	21	5	6.59	46	15	2.62	316.2775	46.2507
1558	21	5	23.78	34	34	11.85	316.3491	34.5700
1559	21	5	28.77	46	17	45.52	316.3699	46.2960
1560	21	5	41.95	34	31	47.12	316.4248	34.5298
1561	21	6	8.57	46	34	18.11	316.5357	46.5717
1562	21	6	35.6	34	42	43.46	316.6483	34.7121
1563	21	6	48.8	34	36	1.24	316.7033	34.6003
1564	21	6	57.68	46	48	3.77	316.7403	46.8010
1565	21	7	41.09	30	40	34.83	316.9212	30.6763

1566	21	8	29.63	47	15	25.37	317.1235	47.2570
1567	21	8	43.98	36	50	34.12	317.1832	36.8428
1568	21	9	18.83	38	58	44.87	317.3285	38.9791
1569	21	10	18.21	31	52	15.26	317.5759	31.8709
1570	21	11	18.99	45	2	46.33	317.8291	45.0462
1571	21	11	33.6	34	22	6.08	317.8900	34.3684
1572	21	13	22.05	37	31	36.26	318.3419	37.5267
1573	21	13	44.62	44	56	29.11	318.4359	44.9414
1574	21	14	36.09	37	30	45.84	318.6504	37.5127
1575	21	15	52.31	35	34	31.64	318.9680	35.5755
1576	21	16	8.11	42	28	24.53	319.0338	42.4735
1577	21	16	26.04	47	8	41.23	319.1085	47.1448
1578	21	16	43.14	44	44	39.1	319.1797	44.7442
1579	21	16	59.44	40	19	56.71	319.2476	40.3324
1580	21	17	10.5	46	47	54.78	319.2937	46.7986
1581	21	18	42.78	43	43	30.58	319.6783	43.7252
1582	21	18	48.53	39	43	58.81	319.7022	39.7330
1583	21	18	58.11	33	29	16.68	319.7421	33.4880
1584	21	19	59.67	39	50	9.93	319.9986	39.8361
1585	21	20	9.22	34	33	39.68	320.0384	34.5610
1586	21	20	39.36	39	53	28.53	320.1640	39.8913
1587	21	20	45.43	30	10	15.82	320.1893	30.1711
1588	21	23	2.58	40	42	24.91	320.7608	40.7069
1589	21	23	15.31	37	26	11.07	320.8138	37.4364
1590	21	23	47.04	30	54	53.25	320.9460	30.9148
1591	21	24	31.23	35	52	6.45	321.1301	35.8685
1592	21	26	42.58	31	54	24.05	321.6774	31.9067
1593	21	27	14.93	46	38	30.79	321.8122	46.6419
1594	21	27	16.67	39	17	19.59	321.8195	39.2888
1595	21	28	10.41	44	11	0.9	322.0434	44.1836
1596	21	28	10.94	43	26	54.37	322.0456	43.4484
1597	21	28	17.22	44	35	21.15	322.0717	44.5892
1598	21	30	15.82	52	53	18.72	322.5659	52.8885
1599	21	30	31.2	46	19	49.61	322.6300	46.3304
1600	21	30	50	44	9	5.12	322.7083	44.1514
1601	21	32	19.95	46	27	46.29	323.0831	46.4629
1602	21	32	26.78	52	43	25.22	323.1116	52.7237
1603	21	32	32.57	38	40	50.73	323.1357	38.6808
1604	21	32	56.73	49	47	54.32	323.2364	49.7984
1605	21	33	29.26	43	7	59.41	323.3719	43.1332
1606	21	34	31.24	47	26	60	323.6302	47.4500
1607	21	35	22.82	47	2	24.12	323.8451	47.0400
1608	21	35	57.9	52	23	42.01	323.9913	52.3950

1609	21	36	47.67	49	32	1.65	324.1986	49.5338
1610	21	38	3.09	50	24	19.78	324.5129	50.4055
1611	21	38	4.56	33	7	27.88	324.5190	33.1244
1612	21	38	8.82	50	35	16.82	324.5367	50.5880
1613	21	38	12.73	33	10	7.04	324.5531	33.1686
1614	21	38	18.82	48	28	13.78	324.5784	48.4705
1615	21	38	19.78	50	50	2.78	324.5824	50.8341
1616	21	38	40.84	49	26	55.94	324.6702	49.4489
1617	21	38	46.99	50	19	14.5	324.6958	50.3207
1618	21	38	48.74	49	27	7.48	324.7031	49.4521
1619	21	39	12.39	50	23	40.81	324.8016	50.3947
1620	21	39	31.69	51	34	35.79	324.8820	51.5766
1621	21	39	34.55	45	45	59.9	324.8940	45.7666
1622	21	39	46.77	51	34	10.28	324.9449	51.5695
1623	21	40	1.77	50	39	45.5	325.0074	50.6626
1624	21	40	27.5	48	22	31.6	325.1146	48.3754
1625	21	40	35.97	49	37	7.44	325.1499	49.6187
1626	21	40	58.71	51	44	38.7	325.2446	51.7441
1627	21	42	14.26	44	34	56.61	325.5594	44.5824
1628	21	42	16.81	46	17	35.81	325.5701	46.2933
1629	21	42	37.82	45	11	38.48	325.6576	45.1940
1630	21	45	18.08	46	34	51.25	326.3253	46.5809
1631	21	46	12.01	45	10	48.88	326.5500	45.1802
1632	21	46	32.88	48	11	55.08	326.6370	48.1986
1633	21	46	56.07	29	4	1.62	326.7336	29.0671
1634	21	48	54.02	48	32	46.77	327.2251	48.5463
1635	21	49	27.51	44	58	10.03	327.3646	44.9695
1636	21	49	30.64	42	20	10.61	327.3777	42.3363
1637	21	50	27.29	41	55	10.42	327.6137	41.9196
1638	21	50	49.48	47	54	45.93	327.7061	47.9128
1639	21	51	19.26	44	27	42.35	327.8302	44.4618
1640	21	52	13.45	43	47	1.35	328.0560	43.7837
1641	21	54	20.88	45	10	13.45	328.5870	45.1704
1642	21	54	36.18	50	36	9.26	328.6507	50.6026
1643	21	55	21.35	52	43	13.85	328.8390	52.7205
1644	21	56	7.54	53	57	42.09	329.0314	53.9617
1645	21	56	34.03	53	22	14.52	329.1418	53.3707
1646	21	56	53.06	53	26	13.41	329.2211	53.4371
1647	21	57	8.57	50	14	35.24	329.2857	50.2431
1648	21	57	12.25	48	36	30.33	329.3010	48.6084
1649	21	57	25.13	50	15	6.88	329.3547	50.2519
1650	21	57	36.92	47	25	22.64	329.4038	47.4230
1651	21	58	1.64	48	20	48.54	329.5068	48.3468

1652	21	59	11.5	55	56	52.59	329.7979	55.9479
1653	21	59	54.25	50	36	31.78	329.9760	50.6088
1654	22	1	49.04	48	51	33.27	330.4544	48.8592
1655	22	2	21.8	50	44	13.61	330.5908	50.7371
1656	22	3	13.69	54	35	10.6	330.8071	54.5863

Nicholson #1656 – 15 x 15 arc minutes

DELPHINUS

#	H	M	S	D	M	S	RA DECIMAL	DEC DECIMAL
1657	20	10	0.78	11	53	33.75	302.5033	11.8927
1658	20	13	23.84	13	27	23.65	303.3493	13.4566
1659	20	16	32.89	12	3	39.78	304.1370	12.0611
1660	20	16	45.56	12	0	43.67	304.1898	12.0121
1661	20	19	17.85	19	12	40.68	304.8244	19.2113
1662	20	19	50.57	19	31	33.72	304.9607	19.5260
1663	20	19	56.71	19	32	15.38	304.9863	19.5376
1664	20	22	15.11	22	44	30.41	305.5630	22.7418
1665	20	23	0	13	17	47.95	305.7500	13.2967
1666	20	23	22.57	21	10	27.47	305.8440	21.1743
1667	20	24	5.53	21	58	10.02	306.0230	21.9694
1668	20	25	19.15	19	54	1.61	306.3298	19.9004
1669	20	25	24.92	19	55	29.94	306.3538	19.9250
1670	20	26	10.8	20	20	18.59	306.5450	20.3385
1671	20	26	41.21	20	41	27.99	306.6717	20.6911
1672	20	27	3.92	18	29	26.77	306.7663	18.4908
1673	20	27	4.22	21	4	31.58	306.7676	21.0754
1674	20	37	35.3	12	5	42.73	309.3971	12.0952
1675	20	37	48.91	9	29	27.78	309.4538	9.4911
1676	20	38	39.51	18	19	24.06	309.6646	18.3234
1677	20	39	9.04	17	52	5.34	309.7877	17.8682
1678	20	39	15.38	10	56	53.09	309.8141	10.9481
1679	20	46	30.42	6	48	33.89	311.6267	6.8094
1680	20	53	9.75	19	55	59.97	313.2906	19.9333
1681	21	6	42.87	12	7	39.83	316.6786	12.1277

DORADO

#	H	M	S	D	M	S	RA DECIMAL	DEC DECIMAL
1682	5	16	15.3	-55	40	48.53	79.0637	-55.6801

DRACO

#	H	M	S	D	M	S	RA DECIMAL	DEC DECIMAL
1683	19	4	42.19	54	56	27.4	286.1758	54.9409
1684	19	22	39.26	60	8	55.75	290.6636	60.1488
1685	19	48	4.76	60	58	40.88	297.0198	60.9780
1686	20	3	26.15	58	44	28.68	300.8589	58.7413

GEMINI

#	H	M	S	D	M	S	RA DECIMAL	DEC DECIMAL
1687	5	59	10.5	27	34	24.76	89.7937	27.5735
1688	6	1	18.25	27	16	52.8	90.3261	27.2813
1689	6	2	54.07	26	1	31.37	90.7253	26.0254
1690	6	4	45.3	25	14	9.84	91.1887	25.2361
1691	6	5	58.46	26	36	35.44	91.4936	26.6098
1692	6	15	58.73	25	33	18.76	93.9947	25.5552
1693	6	19	41.85	21	20	59.25	94.9244	21.3498
1694	6	20	27.38	11	2	55.67	95.1141	11.0488
1695	6	21	27.9	21	8	52.94	95.3662	21.1480
1696	6	21	43.84	16	22	57.62	95.4327	16.3827
1697	6	25	58.74	21	23	5.15	96.4948	21.3848
1698	6	26	17.16	25	33	20.51	96.5715	25.5557
1699	6	26	37.75	27	9	8.32	96.6573	27.1523
1700	6	27	30.03	24	32	0.87	96.8751	24.5336
1701	6	28	28.82	23	21	6.91	97.1201	23.3519
1702	6	29	36.16	19	52	33.71	97.4007	19.8760
1703	6	33	6.6	13	5	43.13	98.2775	13.0953
1704	6	36	3.38	17	52	7.72	99.0141	17.8688
1705	6	37	9.9	16	6	33.73	99.2912	16.1094
1706	6	41	10.57	24	12	32.36	100.2940	24.2090
1707	6	41	21.39	24	13	19.68	100.3391	24.2221
1708	6	41	50.32	20	54	15.2	100.4597	20.9042

1709	6	41	53.91	16	13	56.56	100.4746	16.2324
1710	6	42	8.03	17	9	13.09	100.5335	17.1536
1711	6	42	17.06	14	42	43.64	100.5711	14.7121
1712	6	43	43.89	17	3	38.39	100.9329	17.0607
1713	6	45	12.28	22	43	27.12	101.3012	22.7242
1714	6	45	14.41	13	44	36.02	101.3101	13.7433
1715	6	45	49.39	18	41	28.34	101.4558	18.6912
1716	6	46	16.16	21	31	53	101.5673	21.5314
1717	6	46	43.57	20	53	21.56	101.6816	20.8893
1718	6	47	6.44	14	47	51.19	101.7769	14.7976
1719	6	47	54.99	15	12	9.88	101.9791	15.2027
1720	6	48	46.01	23	48	20.82	102.1917	23.8058
1721	6	48	46.97	30	32	2.31	102.1957	30.5340
1722	6	49	56.9	16	15	23.57	102.4871	16.2565
1723	6	51	13.38	15	11	42.77	102.8058	15.1952
1724	6	51	16.57	15	45	13.39	102.8191	15.7537
1725	6	51	58.65	12	4	55.74	102.9944	12.0821
1726	6	53	5.59	23	51	11.44	103.2733	23.8532
1727	6	54	12.41	24	44	18.5	103.5517	24.7385
1728	6	54	50.11	28	54	48.96	103.7088	28.9136
1729	6	54	52.47	14	1	32.82	103.7186	14.0258
1730	6	58	17.56	14	16	40.81	104.5732	14.2780
1731	6	59	42.72	37	3	50.01	104.9280	37.0639
1732	7	0	18.99	15	23	53.49	105.0791	15.3982
1733	7	1	47.36	15	40	23.19	105.4474	15.6731
1734	7	3	33.58	21	36	0.6	105.8899	21.6002
1735	7	9	56.36	16	34	2.46	107.4848	16.5674
1736	7	11	15.69	20	12	43.31	107.8154	20.2120
1737	7	12	55.57	24	38	49.45	108.2315	24.6471
1738	7	13	10.83	24	39	23.08	108.2951	24.6564
1739	7	21	17.22	13	4	42.12	110.3218	13.0784

HERCULES

#	H	M	S	D	M	S	RA DECIMAL	DEC DECIMAL
1740	17	40	31.34	21	43	53.88	265.1306	21.7316
1741	17	42	25.29	36	4	48.18	265.6054	36.0800
1742	18	5	52.6	21	27	37.52	271.4692	21.4604
1743	18	14	11.52	29	37	25.89	273.5480	29.6239
1744	18	16	49.37	20	37	40.83	274.2057	20.6280
1745	18	20	19.46	21	21	42.99	275.0811	21.3619
1746	18	22	31.68	20	33	42.41	275.6320	20.5618
1747	18	26	20.74	20	28	42.31	276.5864	20.4784
1748	18	31	12.08	25	15	49.63	277.8003	25.2638
1749	18	33	12.94	17	25	59.56	278.3039	17.4332
1750	18	42	14.25	24	52	36.15	280.5594	24.8767
1751	18	43	8.48	20	57	10.1	280.7853	20.9528
1752	18	43	52.54	23	33	24.76	280.9689	23.5569
1753	18	46	4.28	23	1	44.87	281.5178	23.0291
1754	18	46	17.5	22	25	36.78	281.5729	22.4269
1755	18	49	11.84	16	23	37.64	282.2993	16.3938
1756	18	50	53.43	16	34	7.26	282.7226	16.5687
1757	18	51	57.06	17	29	35.6	282.9878	17.4932
1758	18	54	47.53	24	7	38.52	283.6980	24.1274
1759	18	55	17.49	20	0	55.08	283.8229	20.0153

Nicholson #1759 – 15 x 15 arc minutes

HYDRA

#	H	M	S	D	M	S	RA DECIMAL	DEC DECIMAL
1760	8	29	59.35	-10	13	39.54	127.4973	-10.2277
1761	9	5	39.64	-4	51	45.32	136.4151	-4.8626

LACERTA

#	H	M	S	D	M	S	RA DECIMAL	DEC DECIMAL
1762	21	56	52.22	45	19	54.06	329.2176	45.3317
1763	21	57	50.5	52	33	35.24	329.4604	52.5598
1764	21	59	40.02	44	39	51.37	329.9167	44.6643
1765	22	0	4.67	45	1	0.53	330.0195	45.0168
1766	22	1	2.31	45	23	6.34	330.2596	45.3851
1767	22	1	27.45	46	18	23.35	330.3644	46.3065
1768	22	1	42.91	41	43	8.16	330.4288	41.7189
1769	22	3	26.65	42	50	17.99	330.8610	42.8383
1770	22	3	39.59	49	37	55.03	330.9150	49.6320
1771	22	5	8.19	47	30	39.5	331.2841	47.5110
1772	22	6	46.59	51	5	42.03	331.6941	51.0950
1773	22	7	11.34	37	27	17.36	331.7972	37.4548
1774	22	7	41.44	49	38	54.16	331.9227	49.6484
1775	22	8	17.25	51	26	46.21	332.0719	51.4462
1776	22	8	23.3	51	49	12.66	332.0971	51.8202
1777	22	8	30.05	48	52	49.37	332.1252	48.8804
1778	22	10	4.83	50	55	15.05	332.5201	50.9208
1779	22	10	11.47	53	33	18.62	332.5478	53.5552
1780	22	10	15.93	53	28	41.65	332.5664	53.4782
1781	22	10	22.9	50	11	38.13	332.5954	50.1939
1782	22	10	26.26	53	28	32.13	332.6094	53.4756
1783	22	10	26.7	47	57	42.89	332.6113	47.9619
1784	22	10	30.07	54	4	19.91	332.6253	54.0722
1785	22	10	31.71	53	27	1.49	332.6321	53.4504
1786	22	10	34.45	47	54	49.6	332.6435	47.9138
1787	22	10	46.17	53	26	56.42	332.6924	53.4490
1788	22	12	4.34	53	56	47.05	333.0181	53.9464
1789	22	12	10.96	48	59	38.55	333.0456	48.9940
1790	22	12	23.9	51	19	0.24	333.0996	51.3167
1791	22	12	57.26	53	32	32.17	333.2386	53.5423
1792	22	14	2.71	52	22	51.67	333.5113	52.3810

1793	22	14	47	53	0	34.99	333.6959	53.0097
1794	22	14	49.04	52	25	1.04	333.7044	52.4170
1795	22	14	55.1	49	19	28.36	333.7296	49.3245
1796	22	14	56.11	52	58	54.42	333.7338	52.9818
1797	22	15	0.69	52	29	6.03	333.7529	52.4850
1798	22	15	33.82	50	46	4.9	333.8909	50.7680
1799	22	15	39.76	52	55	36.67	333.9157	52.9269
1800	22	15	45.35	50	51	54.74	333.9390	50.8652
1801	22	15	56.01	52	55	33.38	333.9834	52.9259
1802	22	16	27.75	53	28	46.44	334.1156	53.4796
1803	22	16	37.83	52	0	50.33	334.1576	52.0140
1804	22	16	41.43	52	35	35.39	334.1726	52.5932
1805	22	16	46.72	52	32	52.82	334.1947	52.5480
1806	22	17	34.03	53	5	38.33	334.3918	53.0940
1807	22	18	17.35	51	36	54.48	334.5723	51.6151
1808	22	18	31.74	49	36	2.7	334.6322	49.6007
1809	22	19	26.7	51	16	6.78	334.8613	51.2686
1810	22	19	34.81	51	52	56.04	334.8951	51.8822
1811	22	19	36.28	51	15	39.84	334.9012	51.2611
1812	22	19	55.95	49	28	51.59	334.9831	49.4810
1813	22	20	16.13	50	36	47.65	335.0672	50.6132
1814	22	20	20.81	53	8	9.52	335.0867	53.1360
1815	22	20	27.11	50	59	43.82	335.1129	50.9955
1816	22	20	51.71	50	2	4.01	335.2155	50.0344
1817	22	23	28.72	55	15	6.46	335.8697	55.2518
1818	22	23	41.92	53	50	33.08	335.9247	53.8425
1819	22	24	47.52	54	48	18.65	336.1980	54.8052
1820	22	24	52.95	52	7	58.33	336.2206	52.1329
1821	22	26	28.95	53	53	14.92	336.6206	53.8875
1822	22	26	38.13	49	22	11.98	336.6589	49.3700
1823	22	26	58.26	46	1	48.8	336.7427	46.0302
1824	22	29	10.75	45	29	8.65	337.2948	45.4857
1825	22	29	40.89	51	47	20.65	337.4204	51.7891
1826	22	31	22.91	50	40	46.97	337.8455	50.6797
1827	22	32	59.6	48	3	45.11	338.2483	48.0625
1828	22	33	2.9	42	10	4.43	338.2621	42.1679
1829	22	33	44.57	49	26	54.22	338.4357	49.4484
1830	22	35	18.19	51	25	53.78	338.8258	51.4316
1831	22	37	1.47	47	56	3.3	339.2561	47.9343
1832	22	37	15.31	51	17	50.31	339.3138	51.2973
1833	22	37	21.98	54	18	22.71	339.3416	54.3063
1834	22	37	58.6	48	51	6.6	339.4942	48.8518
1835	22	38	0.55	45	0	12.54	339.5023	45.0035

1836	22	40	45.11	48	36	28.22	340.1880	48.6078
1837	22	40	54.22	51	0	27.42	340.2259	51.0076
1838	22	42	16.39	50	27	18.12	340.5683	50.4550
1839	22	42	39.39	43	54	37.41	340.6641	43.9104
1840	22	43	17.44	44	41	11.8	340.8227	44.6866
1841	22	43	36.46	53	24	41.44	340.9019	53.4115
1842	22	43	53.36	47	46	54.62	340.9723	47.7818
1843	22	44	39.69	53	36	44.39	341.1654	53.6123
1844	22	45	1.97	53	5	59.65	341.2582	53.0999
1845	22	45	32.74	41	2	27.22	341.3864	41.0409
1846	22	48	17.64	47	23	5.1	342.0735	47.3848
1847	22	49	28.22	48	36	51.89	342.3676	48.6144
1848	22	50	5.15	49	47	36.64	342.5215	49.7935
1849	22	50	37.15	53	7	6.11	342.6548	53.1184
1850	22	50	41.14	49	31	3.94	342.6714	49.5178
1851	22	51	56.98	53	4	8.9	342.9874	53.0691
1852	22	52	44.51	48	27	44.17	343.1855	48.4623
1853	22	53	21.66	53	16	30.77	343.3402	53.2752
1854	22	54	2.37	49	11	38.03	343.5099	49.1939
1855	22	54	35.36	47	19	2.01	343.6473	47.3172
1856	22	54	42.74	43	33	1.99	343.6781	43.5506
1857	22	54	59.53	51	31	31.87	343.7480	51.5255
1858	22	56	10.37	50	41	3.03	344.0432	50.6842
1859	22	57	57.35	50	32	14.28	344.4890	50.5373
1860	22	58	12.84	48	26	57.16	344.5535	48.4492
1861	22	58	51.76	50	14	30.15	344.7157	50.2417
1862	22	59	19.93	52	28	59.69	344.8330	52.4832
1863	23	2	16.98	51	35	14.37	345.5707	51.5873
1864	23	3	14.89	52	0	57.03	345.8121	52.0158

LEPUS

#	H	M	S	D	M	S	RA DECIMAL	DEC DECIMAL
1865	5	50	0.71	-15	19	5.49	87.5030	-15.3182
1866	6	1	16.11	-27	23	1.76	90.3171	-27.3838

LIBRA

#	H	M	S	D	M	S	RA DECIMAL	DEC DECIMAL
1867	15	24	1.37	-26	56	56.53	231.0057	-26.9490

LUPUS

#	H	M	S	D	M	S	RA DECIMAL	DEC DECIMAL
1868	14	12	40.49	-55	51	30.99	213.1687	-55.8586
1869	14	16	44.83	-46	19	9.11	214.1868	-46.3192
1870	14	21	24.09	-43	18	3.62	215.3504	-43.3010
1871	14	24	51.3	-52	48	7.03	216.2137	-52.8020
1872	14	26	9.33	-53	19	26.22	216.5389	-53.3240
1873	14	38	47.36	-48	12	34.36	219.6973	-48.2095
1874	14	43	46.39	-44	30	11.73	220.9433	-44.5033
1875	14	47	30.49	-52	54	56.72	221.8770	-52.9158
1876	14	50	0.9	-48	38	8.93	222.5037	-48.6358
1877	14	53	30.8	-54	52	24.88	223.3783	-54.8736
1878	14	55	6.14	-43	55	33.49	223.7756	-43.9260
1879	14	56	22.92	-45	9	22.44	224.0955	-45.1562
1880	14	57	23.64	-50	5	10.16	224.3485	-50.0862
1881	15	0	38.78	-44	44	17.93	225.1616	-44.7383
1882	15	1	31.25	-54	58	13.61	225.3802	-54.9704
1883	15	1	59.94	-54	56	48.37	225.4997	-54.9468
1884	15	2	17.35	-49	47	22.72	225.5723	-49.7896
1885	15	2	24.92	-52	50	15.22	225.6038	-52.8376
1886	15	2	42.51	-52	49	58.03	225.6771	-52.8328
1887	15	4	23.07	-49	44	33.68	226.0961	-49.7427
1888	15	6	1.14	-48	26	0.18	226.5047	-48.4334
1889	15	6	22.92	-48	14	13.67	226.5955	-48.2371
1890	15	10	14.68	-44	31	16.83	227.5612	-44.5213
1891	15	10	46.79	-46	4	0.69	227.6950	-46.0669
1892	15	11	57.39	-47	18	58.82	227.9891	-47.3163
1893	15	12	26.75	-49	36	21.85	228.1115	-49.6061
1894	15	12	39.23	-52	29	28.59	228.1635	-52.4913
1895	15	13	24.81	-50	24	50.04	228.3534	-50.4139
1896	15	13	54.56	-52	21	40.04	228.4773	-52.3611
1897	15	13	56.52	-54	36	12.49	228.4855	-54.6035
1898	15	15	43.8	-42	41	39.33	228.9325	-42.6943
1899	15	16	20.77	-41	23	59.8	229.0865	-41.3999

1900	15	16	45.57	-51	54	39.69	229.1899	-51.9110
1901	15	22	33.33	-46	43	22.06	230.6389	-46.7228
1902	15	24	1.37	-41	59	46.72	231.0057	-41.9963
1903	15	24	33.28	-40	26	38.67	231.1387	-40.4441
1904	15	39	49.8	-45	34	41.95	234.9575	-45.5783
1905	15	39	55.55	-46	22	51.58	234.9815	-46.3810
1906	15	40	29.36	-40	11	56.44	235.1223	-40.1990
1907	15	41	23.65	-42	10	34.63	235.3485	-42.1763
1908	15	42	6.44	-45	30	22.02	235.5268	-45.5061
1909	15	42	16.32	-45	30	56.01	235.5680	-45.5156
1910	15	52	24.17	-40	58	4.24	238.1007	-40.9678

Nicholson #1909 – 15 x 15 arc minutes

LYNX

#	H	M	S	D	M	S	RA DECIMAL	DEC DECIMAL
1911	7	48	46.65	36	40	16.44	117.1944	36.6712

LYRA

#	H	M	S	D	M	S	RA DECIMAL	DEC DECIMAL
1912	18	51	37.82	37	28	8.56	282.9076	37.4690
1913	18	55	14.12	27	47	23.95	283.8088	27.7900
1914	18	55	25.22	40	12	4.91	283.8551	40.2014
1915	18	56	40.02	33	7	41.05	284.1668	33.1281
1916	18	56	52.65	24	23	17.51	284.2194	24.3882
1917	18	57	1.72	31	24	8.45	284.2572	31.4023
1918	18	57	2.06	26	8	39.33	284.2586	26.1443
1919	18	57	7.54	34	51	19.76	284.2814	34.8555
1920	18	59	9.53	39	55	2.67	284.7897	39.9174
1921	19	1	54.03	30	0	3.58	285.4751	30.0010
1922	19	2	59.15	27	17	6.35	285.7464	27.2851
1923	19	4	11.41	40	30	40.13	286.0476	40.5111
1924	19	4	31.47	36	39	20.76	286.1311	36.6558
1925	19	4	42.99	37	17	19.96	286.1791	37.2889
1926	19	4	56	28	28	27.09	286.2333	28.4742
1927	19	5	1.64	29	5	34.59	286.2568	29.0929
1928	19	7	38.39	28	32	48.96	286.9099	28.5469
1929	19	7	50.02	29	27	12.68	286.9584	29.4535
1930	19	8	9.91	26	14	15.41	287.0413	26.2376
1931	19	8	42.01	37	56	20.01	287.1751	37.9389
1932	19	10	38.93	42	15	36.89	287.6622	42.2602
1933	19	10	46.87	31	19	17.83	287.6953	31.3216
1934	19	10	52.23	35	32	28.31	287.7176	35.5412
1935	19	11	33.6	29	0	18.13	287.8900	29.0050
1936	19	11	38.21	31	29	3.67	287.9092	31.4844
1937	19	12	11.23	28	32	41.13	288.0468	28.5448
1938	19	14	33.19	29	34	23.23	288.6383	29.5731
1939	19	17	47.31	27	54	38.11	289.4471	27.9106
1940	19	18	50.15	39	40	56.34	289.7090	39.6823
1941	19	19	43.57	29	1	52.48	289.9315	29.0312
1942	19	20	15.54	29	12	1.42	290.0647	29.2004
1943	19	21	20.12	33	13	54.42	290.3338	33.2318

1944	19	21	38.36	30	21	11.03	290.4098	30.3531
1945	19	21	49.78	32	59	31.18	290.4574	32.9920
1946	19	22	35.03	40	3	47.27	290.6460	40.0631
1947	19	24	29.34	28	31	43.39	291.1222	28.5287
1948	19	24	37.07	43	1	19.75	291.1544	43.0222
1949	19	25	0.02	31	24	49.19	291.2501	31.4137
1950	19	25	19.2	44	32	48.8	291.3300	44.5469
1951	19	25	28.78	38	21	27.22	291.3699	38.3576
1952	19	25	32.31	30	54	51.04	291.3846	30.9142
1953	19	25	32.51	38	23	6.19	291.3854	38.3851
1954	19	26	35.37	31	17	25.12	291.6474	31.2903
1955	19	26	46.52	29	28	52.87	291.6938	29.4814
1956	19	27	7.63	32	15	58.02	291.7818	32.2661
1957	19	27	8.28	38	42	11.4	291.7845	38.7032
1958	19	28	46.36	29	25	39.56	292.1932	29.4277
1959	19	29	19.7	32	15	8.63	292.3321	32.2524
1960	19	30	24.8	31	17	13.96	292.6033	31.2872
1961	19	30	36.5	31	13	24.78	292.6521	31.2236
1962	19	30	44.17	31	10	11.75	292.6841	31.1699
1963	19	31	4.29	38	11	12.81	292.7679	38.1869
1964	19	31	20.45	30	9	26.62	292.8352	30.1574
1965	19	31	25.52	31	44	23.7	292.8563	31.7399
1966	19	33	31.26	31	31	3.35	293.3803	31.5176

MONOCEROS

#	H	M	S	D	M	S	RA DECIMAL	DEC DECIMAL
1967	6	11	1.38	-9	13	56.66	92.7558	-9.2324
1968	6	20	11.68	10	24	58.8	95.0487	10.4163
1969	6	21	19.25	8	24	55.82	95.3302	8.4155
1970	6	22	10.71	9	1	27.17	95.5446	9.0242
1971	6	24	32.99	7	30	7.27	96.1375	7.5020
1972	6	27	31.99	8	30	32.87	96.8833	8.5091
1973	6	29	39.84	-2	13	28.2	97.4160	-2.2245
1974	6	30	4.51	7	20	22.04	97.5188	7.3395
1975	6	30	56.38	0	57	48.6	97.7349	-0.9635
1976	6	32	21.87	-5	46	8.77	98.0911	-5.7691
1977	6	34	45.83	-1	7	4.04	98.6909	-1.1178
1978	6	35	3.82	-4	38	29.48	98.7659	-4.6415
1979	6	35	25.97	-11	59	20.64	98.8582	-11.9891
1980	6	37	4.83	11	29	48.3	99.2701	11.4967
1981	6	39	24.52	3	57	15.17	99.8521	3.9542
1982	6	39	56.51	-4	33	39.96	99.9855	-4.5611
1983	6	41	4.04	2	14	24.76	100.2668	2.2402
1984	6	41	5.88	9	22	55.63	100.2745	9.3821
1985	6	41	15.99	4	20	36.35	100.3166	4.3434
1986	6	41	27.23	-5	51	16.99	100.3635	-5.8547
1987	6	41	31.27	-7	48	21.58	100.3803	-7.8060
1988	6	41	43.19	-9	23	33.39	100.4300	-9.3926
1989	6	42	8.79	-11	32	37.59	100.5366	-11.5438
1990	6	42	30.97	5	10	5.82	100.6290	5.1683
1991	6	42	36.8	-1	29	7.62	100.6533	-1.4855
1992	6	45	4.52	10	47	3.12	101.2688	10.7842
1993	6	45	33.67	8	2	20.83	101.3903	8.0391
1994	6	45	35.83	3	47	35.68	101.3993	3.7932
1995	6	45	40.06	-5	22	25.62	101.4169	-5.3738
1996	6	45	56.16	-6	46	1.79	101.4840	-6.7672
1997	6	46	37.72	-8	49	41.92	101.6571	-8.8283
1998	6	48	14.39	-5	10	29.03	102.0599	-5.1747
1999	6	49	50.31	-9	29	25.98	102.4596	-9.4905
2000	6	50	12.66	0	47	9.64	102.5528	0.7860
2001	6	50	23.62	-4	59	26.73	102.5984	-4.9908
2002	6	50	56.03	-2	52	33.45	102.7335	-2.8760
2003	6	51	59.81	-9	53	55.93	102.9992	-9.8989

2004	6	52	3.63	-4	8	30.69	103.0151	-4.1419
2005	6	52	19.81	-6	28	8.55	103.0825	-6.4690
2006	6	53	5.4	8	41	26.24	103.2725	8.6906
2007	6	53	12.05	-3	45	39.51	103.3002	-3.7610
2008	6	54	30.44	-6	23	21.31	103.6268	-6.3893
2009	6	55	21.71	7	18	57.81	103.8404	7.3161
2010	6	55	22.86	8	21	29.47	103.8453	8.3582
2011	6	55	26.37	7	16	13.81	103.8599	7.2705
2012	6	55	34.06	0	57	25.01	103.8919	-0.9569
2013	6	58	11.21	5	46	21.72	104.5467	5.7727
2014	6	59	48.07	12	17	31.82	104.9503	12.2922
2015	7	0	39.79	12	29	35.3	105.1658	12.4931
2016	7	2	2.23	10	43	24.52	105.5093	10.7235
2017	7	2	22.66	-6	15	19.34	105.5944	-6.2554
2018	7	3	15.68	-3	20	35.51	105.8153	-3.3432
2019	7	3	21.35	-2	6	37.06	105.8390	-2.1103
2020	7	3	25.91	-3	19	6.23	105.8580	-3.3184
2021	7	3	34.51	-7	50	4.64	105.8938	-7.8346
2022	7	4	2.13	-2	11	3.67	106.0089	-2.1844
2023	7	4	8.88	0	10	11.21	106.0370	0.1698
2024	7	4	12.73	-3	15	13.39	106.0530	-3.2537
2025	7	4	16.97	-2	7	42.98	106.0707	-2.1286
2026	7	4	21	-3	13	37.87	106.0875	-3.2272
2027	7	4	46.21	-8	40	11.85	106.1925	-8.6700
2028	7	5	28.91	-3	20	39.06	106.3704	-3.3442
2029	7	5	30.17	5	22	25.05	106.3757	5.3736
2030	7	5	31.31	1	49	19.26	106.3805	1.8220
2031	7	5	32.85	-7	15	41.35	106.3869	-7.2615
2032	7	5	32.93	-3	20	57.9	106.3872	-3.3494
2033	7	5	40.25	0	20	56.79	106.4177	-0.3491
2034	7	5	49.89	-4	52	47.32	106.4579	-4.8798
2035	7	6	8.31	-9	26	53.63	106.5346	-9.4482
2036	7	6	9.59	-1	32	1.02	106.5399	-1.5336
2037	7	6	15.79	-2	23	45.7	106.5658	-2.3960
2038	7	6	17.01	10	16	1.63	106.5709	10.2671
2039	7	6	25.73	7	3	28.72	106.6072	7.0580
2040	7	6	34.6	-4	28	24.35	106.6442	-4.4734
2041	7	6	57.19	7	12	0.98	106.7383	7.2003
2042	7	7	8.31	-6	38	23.8	106.7846	-6.6399
2043	7	7	49.33	-2	50	22.43	106.9555	-2.8396
2044	7	8	13.04	-3	43	53.69	107.0544	-3.7316
2045	7	8	19.46	-9	28	43.67	107.0811	-9.4788
2046	7	9	19.33	-7	5	16.68	107.3305	-7.0880

2047	7	9	26.65	-7	3	0.59	107.3610	-7.0502
2048	7	9	38.68	-6	0	55.47	107.4112	-6.0154
2049	7	9	51.1	-7	38	31.39	107.4629	-7.6421
2050	7	10	41.54	-4	6	28.91	107.6731	-4.1080
2051	7	10	49.31	4	19	43.1	107.7055	4.3286
2052	7	11	3.99	-2	26	55.91	107.7666	-2.4489
2053	7	11	24.9	-3	31	30.06	107.8538	-3.5250
2054	7	11	53.34	-7	19	50.12	107.9723	-7.3306
2055	7	11	56.5	-4	40	56.37	107.9854	-4.6823
2056	7	12	20.91	-5	7	17.51	108.0871	-5.1215
2057	7	12	27.32	-5	3	55.28	108.1138	-5.0654
2058	7	12	27.53	-2	28	15.93	108.1147	-2.4711
2059	7	12	30.81	-4	12	0.97	108.1284	-4.2003
2060	7	12	43.52	-2	19	32.68	108.1813	-2.3257
2061	7	12	54.07	-11	19	1.37	108.2253	-11.3170
2062	7	12	57.49	1	8	23.41	108.2396	1.1398
2063	7	13	7.1	-2	59	30.12	108.2796	-2.9917
2064	7	13	44.55	-6	53	46.25	108.4356	-6.8962
2065	7	13	56.61	-3	44	31.18	108.4859	-3.7420
2066	7	14	42.32	-6	57	6.05	108.6763	-6.9517
2067	7	15	46.19	-7	26	38.01	108.9424	-7.4439
2068	7	15	48.42	-1	2	35.38	108.9517	-1.0432
2069	7	15	53.29	-7	28	4.83	108.9720	-7.4680
2070	7	15	55.65	-9	24	4.12	108.9819	-9.4011
2071	7	16	2.57	-2	26	4.19	109.0107	-2.4345
2072	7	16	2.99	-1	11	55.8	109.0124	-1.1988
2073	7	16	9.53	-6	34	38.68	109.0397	-6.5774
2074	7	16	11.23	-2	15	53.48	109.0468	-2.2649
2075	7	16	15.61	-6	35	0.18	109.0650	-6.5834
2076	7	16	15.69	-2	25	39.17	109.0654	-2.4275
2077	7	16	15.74	-6	34	59.43	109.0656	-6.5832
2078	7	16	24.84	-7	54	42.78	109.1035	-7.9119
2079	7	16	25.47	-6	34	9.87	109.1061	-6.5694
2080	7	16	42.27	-2	6	39.3	109.1761	-2.1109
2081	7	17	10.24	-1	32	14.23	109.2926	-1.5373
2082	7	17	17.2	-10	18	0.47	109.3217	-10.3001
2083	7	17	36.12	-4	44	9.16	109.4005	-4.7359
2084	7	17	39.46	-4	47	12.74	109.4144	-4.7869
2085	7	17	45.49	-5	28	39.11	109.4396	-5.4775
2086	7	18	18.23	-7	45	2.27	109.5759	-7.7506
2087	7	21	10.01	-5	30	47.49	110.2917	-5.5132
2088	7	21	15.58	-5	35	33.73	110.3149	-5.5927
2089	7	22	36.32	-8	4	30.51	110.6513	-8.0751

2090	7	23	15.47	-10	4	6.16	110.8144	-10.0684
2091	7	24	59.12	-8	5	16.54	111.2463	-8.0879
2092	7	25	17.34	-4	24	20.35	111.3223	-4.4057
2093	7	25	19.36	-8	34	54.19	111.3307	-8.5817
2094	7	25	44.56	-9	4	41.17	111.4357	-9.0781
2095	7	26	48.38	-10	25	36.75	111.7016	-10.4269
2096	7	26	52.27	-8	43	7.55	111.7178	-8.7188
2097	7	31	15.59	-5	0	52.57	112.8150	-5.0146
2098	7	32	27.39	-9	27	29.19	113.1141	-9.4581
2099	7	32	49.31	-5	59	14.04	113.2055	-5.9872
2100	7	34	58.77	-4	45	27.97	113.7449	-4.7578
2101	7	44	18.39	-7	49	28.39	116.0766	-7.8246
2102	7	45	32.33	0	24	4.46	116.3847	-0.4012
2103	7	45	40.29	-7	27	57.75	116.4179	-7.4660
2104	7	45	56.48	-11	2	12.6	116.4853	-11.0368
2105	7	46	56.73	-7	13	19.07	116.7364	-7.2220
2106	7	49	53.27	-10	13	51.66	117.4720	-10.2310
2107	7	50	47.87	-11	50	42.49	117.6995	-11.8451
2108	7	54	51.06	-10	21	2.57	118.7128	-10.3507
2109	8	0	4.25	-8	19	53.1	120.0177	-8.3314

Nicholson #2109 – 15 x 15 arc minutes

MUSCA

#	H	M	S	D	M	S	RA DECIMAL	DEC DECIMAL
2110	11	19	52.75	-70	37	6.45	169.9698	-70.6185
2111	11	32	6.06	-67	2	18.01	173.0253	-67.0383
2112	11	34	4.06	-67	25	20.47	173.5169	-67.4224
2113	11	35	13.22	-66	2	11.76	173.8051	-66.0366
2114	11	35	33.5	-67	55	58.63	173.8896	-67.9330
2115	11	36	56.64	-66	47	3.77	174.2360	-66.7844
2116	11	37	22.16	-67	34	44.6	174.3423	-67.5791
2117	11	40	35.82	-65	8	40.52	175.1492	-65.1446
2118	11	40	38.95	-65	6	17.74	175.1623	-65.1049
2119	11	44	15.06	-64	52	18.92	176.0628	-64.8719
2120	11	47	43.55	-67	51	24.3	176.9315	-67.8567
2121	11	47	52.91	-65	59	37.02	176.9705	-65.9936
2122	11	49	53.13	-66	14	39.77	177.4714	-66.2444
2123	11	52	34.3	-72	49	3.91	178.1429	-72.8178
2124	11	54	27.59	-69	58	21.26	178.6150	-69.9726
2125	11	58	13.72	-68	32	37.62	179.5572	-68.5438
2126	11	58	25.26	-68	31	39.2	179.6053	-68.5276
2127	12	10	23.29	-65	42	22.7	182.5970	-65.7063
2128	12	14	10.41	-66	49	54.59	183.5434	-66.8318
2129	12	14	19.81	-66	54	28.1	183.5825	-66.9078
2130	12	17	15.99	-65	18	7.5	184.3166	-65.3021
2131	12	17	38.01	-65	38	44	184.4084	-65.6456
2132	12	17	46.63	-70	6	1.26	184.4443	-70.1004
2133	12	20	51.43	-66	17	20.34	185.2143	-66.2890
2134	12	23	55.8	-65	51	0.22	185.9825	-65.8501
2135	12	25	5.1	-65	0	56.55	186.2713	-65.0157
2136	12	25	23.23	-65	38	54.44	186.3468	-65.6485
2137	12	27	0.34	-65	13	7.35	186.7514	-65.2187
2138	12	33	46.22	-68	52	51.35	188.4426	-68.8809
2139	12	35	9.12	-69	43	25.01	188.7880	-69.7236
2140	12	35	48.29	-66	44	1.33	188.9512	-66.7337
2141	12	47	22.61	-75	31	32.34	191.8442	-75.5257
2142	12	50	16.9	-65	19	13.81	192.5704	-65.3205
2143	12	54	37.88	-72	23	5.37	193.6578	-72.3848
2144	12	55	39.69	-69	0	13.37	193.9154	-69.0037
2145	12	58	1.64	-68	10	11.69	194.5068	-68.1699
2146	13	0	53.43	-66	28	53.82	195.2226	-66.4816
2147	13	1	16.38	-68	36	41.82	195.3182	-68.6116
2148	13	1	37.39	-72	24	31.85	195.4058	-72.4088

2149	13	2	45.32	-69	39	57.15	195.6888	-69.6659
2150	13	3	0.61	-65	43	48.06	195.7525	-65.7300
2151	13	3	33.19	-65	44	44.2	195.8883	-65.7456
2152	13	5	56.58	-72	33	23.85	196.4857	-72.5566
2153	13	6	48.39	-66	48	42.6	196.7016	-66.8118
2154	13	7	2.92	-65	3	47.24	196.7622	-65.0631
2155	13	7	33.57	-70	27	29.21	196.8899	-70.4581
2156	13	8	32.99	-64	38	3.12	197.1374	-64.6342
2157	13	9	4.24	-64	33	12.14	197.2677	-64.5534
2158	13	9	28.97	-66	12	51.8	197.3707	-66.2144
2159	13	10	20.86	-64	46	20.54	197.5869	-64.7724
2160	13	11	46.81	-71	10	50.68	197.9450	-71.1807
2161	13	13	13.32	-68	27	53.96	198.3055	-68.4650
2162	13	24	58.72	-67	19	0.37	201.2447	-67.3168
2163	13	25	39.75	-68	57	8.99	201.4156	-68.9525
2164	13	34	4.21	-66	52	28.22	203.5175	-66.8745
2165	13	34	18.71	-66	50	2.33	203.5780	-66.8340
2166	13	34	37.56	-66	55	21.25	203.6565	-66.9226
2167	13	35	52.96	-65	9	1.66	203.9706	-65.1505
2168	13	39	50.42	-66	41	25.4	204.9601	-66.6904
2169	13	40	15.94	-66	33	18.85	205.0664	-66.5552

Nicholson #2165 – 15 x 15 arc minutes

NORMA

#	H	M	S	D	M	S	RA DECIMAL	DEC DECIMAL
2170	15	16	29.34	-54	35	10	229.1222	-54.5861
2171	15	19	19.92	-56	3	3.07	229.8330	-56.0509
2172	15	19	56.25	-53	16	57.73	229.9844	-53.2827
2173	15	24	56.25	-48	31	14.08	231.2344	-48.5206
2174	15	28	41.24	-54	0	41.37	232.1719	-54.0115
2175	15	30	50.55	-52	57	17.08	232.7106	-52.9547
2176	15	31	24.48	-50	5	54.3	232.8520	-50.0984
2177	15	32	11.95	-49	7	37.11	233.0498	-49.1270
2178	15	33	24.41	-52	59	38.7	233.3517	-52.9941
2179	15	33	36.17	-53	2	34.19	233.4007	-53.0428
2180	15	34	9.56	-52	28	46.62	233.5398	-52.4796
2181	15	35	28.7	-52	7	18.61	233.8696	-52.1218
2182	15	36	26.42	-52	57	31.43	234.1101	-52.9587
2183	15	36	54	-54	49	10.44	234.2250	-54.8196
2184	15	37	41.1	-49	29	31.73	234.4213	-49.4921
2185	15	37	44.75	-53	50	59.45	234.4364	-53.8498
2186	15	38	13.02	-49	6	22.39	234.5543	-49.1062
2187	15	38	20.37	-49	4	21.24	234.5849	-49.0726
2188	15	38	26.13	-53	1	57.68	234.6089	-53.0327
2189	15	38	38.32	-53	19	27.55	234.6597	-53.3243
2190	15	39	30.45	-55	21	12.09	234.8769	-55.3534
2191	15	40	3.51	-49	45	47.71	235.0146	-49.7633
2192	15	40	23.78	-48	22	13.74	235.0991	-48.3705
2193	15	41	39.26	-53	0	47.66	235.4136	-53.0132
2194	15	41	42.67	-56	6	0.31	235.4278	-56.1001
2195	15	43	0.68	-58	52	36.73	235.7528	-58.8769
2196	15	44	8.94	-51	28	51.2	236.0372	-51.4809
2197	15	44	23	-51	26	32.63	236.0958	-51.4424
2198	15	45	37.55	-58	16	28.09	236.4065	-58.2745
2199	15	45	45.52	-46	19	56.42	236.4397	-46.3323
2200	15	46	59.07	-47	12	9.95	236.7461	-47.2028
2201	15	48	10.92	-56	38	30.91	237.0455	-56.6419
2202	15	48	27.62	-59	6	9.46	237.1151	-59.1026
2203	15	49	26.88	-54	34	11.24	237.3620	-54.5698
2204	15	52	47.04	-53	16	46.39	238.1960	-53.2796
2205	15	52	53.84	-50	12	10.94	238.2243	-50.2030
2206	15	53	23.77	-47	51	33.21	238.3490	-47.8592
2207	15	54	8.02	-53	19	40.24	238.5334	-53.3278
2208	15	54	50.67	-50	20	6.07	238.7111	-50.3350

2209	15	55	39.19	-49	51	28.56	238.9133	-49.8579
2210	15	56	34.08	-54	54	14.41	239.1420	-54.9040
2211	15	56	57.55	-49	40	35.54	239.2398	-49.6765
2212	15	57	47.42	-54	15	48.46	239.4476	-54.2635
2213	15	58	32.99	-49	38	54.74	239.6375	-49.6485
2214	15	58	39.06	-58	36	24.57	239.6627	-58.6068
2215	15	59	0.62	-56	56	19.3	239.7526	-56.9387
2216	16	0	0.87	-58	17	26.82	240.0036	-58.2908
2217	16	0	3.28	-50	26	47.17	240.0137	-50.4464
2218	16	0	37.35	-55	32	34.44	240.1556	-55.5429
2219	16	4	23.12	-53	14	37.92	241.0963	-53.2439
2220	16	5	34.85	-53	24	43.31	241.3952	-53.4120
2221	16	7	3.24	-47	14	21.43	241.7635	-47.2393
2222	16	7	7.21	-53	13	33.6	241.7800	-53.2260
2223	16	7	9.72	-47	14	51.31	241.7905	-47.2476
2224	16	8	9.19	-58	28	52.1	242.0383	-58.4811
2225	16	9	17.89	-58	9	42.99	242.3245	-58.1619
2226	16	9	54.33	-52	46	57.72	242.4764	-52.7827
2227	16	11	18.5	-53	45	36.43	242.8271	-53.7601
2228	16	11	28.21	-57	57	5.3	242.8675	-57.9515
2229	16	11	35.73	-57	55	26.3	242.8989	-57.9240
2230	16	11	44.25	-56	5	38.64	242.9344	-56.0941
2231	16	12	40.74	-52	56	57.51	243.1697	-52.9493
2232	16	12	55.62	-58	39	5.45	243.2317	-58.6515
2233	16	12	56.33	-53	38	37.01	243.2347	-53.6436
2234	16	12	57.72	-58	41	46.1	243.2405	-58.6961
2235	16	13	1.02	-53	36	33.31	243.2543	-53.6093
2236	16	13	58.57	-54	54	56.83	243.4940	-54.9158
2237	16	14	2.83	-54	56	58.57	243.5118	-54.9496
2238	16	14	20.91	-53	29	12.72	243.5871	-53.4869
2239	16	14	33.93	-53	33	35.26	243.6414	-53.5598
2240	16	14	49.81	-53	37	29.98	243.7076	-53.6250
2241	16	15	1.82	-56	30	17.45	243.7576	-56.5048
2242	16	15	8	-49	40	17.48	243.7833	-49.6715
2243	16	16	48.29	-58	22	2.43	244.2012	-58.3673
2244	16	16	56.13	-52	32	22.63	244.2339	-52.5396
2245	16	17	35.4	-50	57	43.23	244.3975	-50.9620
2246	16	17	38.56	-54	11	12.78	244.4107	-54.1869
2247	16	17	41.02	-51	2	7.99	244.4209	-51.0356
2248	16	17	51.98	-50	54	54.56	244.4666	-50.9152
2249	16	17	54.02	-54	14	9.69	244.4751	-54.2360
2250	16	18	6.49	-50	57	38.59	244.5270	-50.9607
2251	16	18	36.06	-51	24	28.52	244.6502	-51.4079

2252	16	18	38.45	-53	57	46.3	244.6602	-53.9629
2253	16	18	40.69	-51	12	20.42	244.6695	-51.2057
2254	16	18	41.09	-53	59	42.37	244.6712	-53.9951
2255	16	18	42.36	-58	54	47.17	244.6765	-58.9131
2256	16	18	43.73	-51	27	57.8	244.6822	-51.4661
2257	16	18	48.1	-54	17	25	244.7004	-54.2903
2258	16	18	52.74	-53	41	45.48	244.7198	-53.6960
2259	16	18	53.05	-51	24	28.13	244.7210	-51.4078
2260	16	18	56.96	-53	43	29.83	244.7374	-53.7250
2261	16	19	8.34	-51	5	38.54	244.7847	-51.0940
2262	16	19	10.91	-58	48	19.73	244.7954	-58.8055
2263	16	19	15.39	-53	39	46.92	244.8141	-53.6630
2264	16	19	18.7	-52	26	9.73	244.8279	-52.4360
2265	16	19	21.72	-54	24	22.34	244.8405	-54.4062
2266	16	19	21.88	-53	29	22.31	244.8412	-53.4895
2267	16	19	22.67	-52	22	31.34	244.8444	-52.3754
2268	16	19	28.19	-51	9	17.54	244.8674	-51.1549
2269	16	19	29.24	-51	51	16.48	244.8718	-51.8546
2270	16	19	29.25	-54	18	57.38	244.8719	-54.3159
2271	16	19	44.85	-53	32	58.04	244.9369	-53.5495
2272	16	19	46.64	-53	1	11.81	244.9443	-53.0199
2273	16	19	59.42	-53	37	7.03	244.9976	-53.6186
2274	16	20	18.13	-53	19	55.54	245.0755	-53.3321
2275	16	20	25.19	-53	20	41.54	245.1049	-53.3449
2276	16	20	53.34	-53	12	34.31	245.2222	-53.2095
2277	16	20	56.17	-57	12	8.82	245.2341	-57.2024
2278	16	21	8.38	-53	27	2.89	245.2849	-53.4508
2279	16	21	12.45	-52	36	49.5	245.3019	-52.6138
2280	16	21	14.1	-57	15	59.94	245.3087	-57.2667
2281	16	21	24.54	-52	27	19.29	245.3523	-52.4554
2282	16	21	26.28	-53	35	17.17	245.3595	-53.5881
2283	16	21	26.95	-52	27	28.46	245.3623	-52.4579
2284	16	21	30.94	-53	4	44.05	245.3789	-53.0789
2285	16	21	34.42	-52	34	29.79	245.3934	-52.5749
2286	16	21	36.02	-52	53	53.27	245.4001	-52.8981
2287	16	21	38.35	-53	37	55.43	245.4098	-53.6321
2288	16	21	38.95	-52	29	32.27	245.4123	-52.4923
2289	16	21	49.84	-53	10	19.05	245.4577	-53.1720
2290	16	22	1.66	-52	36	28.42	245.5069	-52.6079
2291	16	22	9.85	-53	23	10.06	245.5410	-53.3861
2292	16	22	15.56	-52	42	9.57	245.5648	-52.7027
2293	16	22	17	-53	25	26.86	245.5708	-53.4241
2294	16	22	20.5	-52	38	2.66	245.5854	-52.6341

2295	16	22	21.87	-49	37	42.66	245.5911	-49.6285
2296	16	22	29.74	-52	49	12.32	245.6239	-52.8201
2297	16	22	34.25	-57	41	44.23	245.6427	-57.6956
2298	16	22	44.45	-52	45	5.41	245.6852	-52.7515
2299	16	22	45.4	-53	26	35.69	245.6892	-53.4432
2300	16	22	47.81	-52	45	55.84	245.6992	-52.7655
2301	16	23	5.17	-53	31	45.56	245.7716	-53.5293
2302	16	23	48.76	-46	36	45.05	245.9532	-46.6125
2303	16	24	23.32	-52	34	44.36	246.0972	-52.5790
2304	16	24	37.39	-53	25	45.15	246.1558	-53.4292
2305	16	24	48.93	-52	33	41.41	246.2039	-52.5615
2306	16	24	50.48	-54	41	2.83	246.2103	-54.6841
2307	16	24	57.47	-52	33	48.94	246.2395	-52.5636
2308	16	25	0.5	-55	37	43.48	246.2521	-55.6287
2309	16	25	13.07	-52	38	27.36	246.3045	-52.6409
2310	16	25	14.67	-47	27	24.78	246.3111	-47.4569
2311	16	25	27.63	-52	35	19.22	246.3651	-52.5887
2312	16	26	37.69	-54	42	42.21	246.6570	-54.7117
2313	16	26	52.5	-52	37	27.81	246.7188	-52.6244
2314	16	27	13.15	-46	54	42.06	246.8048	-46.9117
2315	16	27	17.08	-54	4	45.72	246.8212	-54.0794
2316	16	27	26.81	-43	48	32.77	246.8617	-43.8091
2317	16	27	31.16	-54	4	5.79	246.8798	-54.0683
2318	16	28	14.47	-53	39	33.31	247.0603	-53.6593
2319	16	28	31.87	-50	39	11.76	247.1328	-50.6533
2320	16	28	37.33	-54	36	27.2	247.1556	-54.6076
2321	16	28	38.44	-52	9	21.11	247.1602	-52.1559
2322	16	28	50.46	-52	5	44.51	247.2103	-52.0957
2323	16	29	6.19	-57	25	49.97	247.2758	-57.4305
2324	16	29	11.43	-56	46	4.67	247.2976	-56.7680
2325	16	30	18.45	-53	9	40.36	247.5769	-53.1612
2326	16	31	2.96	-52	3	11.16	247.7623	-52.0531
2327	16	31	19.22	-51	54	11.23	247.8301	-51.9031
2328	16	33	0.03	-57	3	36.63	248.2501	-57.0602
2329	16	33	31.01	-51	33	18.06	248.3792	-51.5550
2330	16	33	37.98	-56	31	27.58	248.4083	-56.5243
2331	16	33	43.49	-51	33	52.75	248.4312	-51.5647
2332	16	33	45.72	-56	35	27.55	248.4405	-56.5910
2333	16	34	2.22	-51	38	2.74	248.5093	-51.6341
2334	16	34	18.82	-51	42	38.16	248.5784	-51.7106
2335	16	34	39.85	-52	11	56.93	248.6660	-52.1991
2336	16	35	16.46	-53	8	27.48	248.8186	-53.1410
2337	16	35	23.37	-53	37	1.84	248.8474	-53.6172

2338	16	35	30.04	-53	37	1.26	248.8752	-53.6170
2339	16	35	36.75	-53	40	28.18	248.9031	-53.6745
2340	16	36	15.28	-57	14	26.82	249.0637	-57.2408
2341	16	36	39.47	-52	26	27.08	249.1644	-52.4409
2342	16	37	20.81	-53	19	24.97	249.3367	-53.3236
2343	16	38	55.23	-53	32	9.83	249.7301	-53.5361
2344	16	39	13.52	-52	26	0.18	249.8063	-52.4334
2345	16	40	19.43	-53	5	31.98	250.0810	-53.0922
2346	16	43	48.04	-52	0	49.83	250.9502	-52.0138

Nicholson #2333 – 15 x 15 arc minutes

OPHIUCUS

#	H	M	S	D	M	S	RA DECIMAL	DEC DECIMAL
2347	16	53	15.49	-29	55	54.43	253.3145	-29.9318
2348	16	53	25.74	-29	56	34.49	253.3572	-29.9429
2349	16	54	16.26	-26	46	37.11	253.5678	-26.7770
2350	16	54	44.54	-28	19	3.96	253.6856	-28.3178
2351	16	55	28.75	-8	20	11.3	253.8698	-8.3365
2352	16	56	28.07	-31	27	34.09	254.1170	-31.4595
2353	17	0	48.81	-28	37	55.17	255.2034	-28.6320
2354	17	1	27.84	-26	46	12.07	255.3660	-26.7700
2355	17	1	35.94	-26	47	30.67	255.3998	-26.7919
2356	17	7	40.14	-30	35	53.18	256.9173	-30.5981
2357	17	8	17.19	-18	49	29.07	257.0716	-18.8247
2358	17	9	0.88	-33	5	3.57	257.2536	-33.0843
2359	17	10	36.42	-25	27	25.58	257.6517	-25.4571
2360	17	11	40.73	-30	42	28.62	257.9197	-30.7080
2361	17	15	51.33	-18	12	27.44	258.9639	-18.2076
2362	17	18	27.51	-19	47	59.97	259.6146	-19.8000
2363	17	23	59.15	-31	27	57.66	260.9965	-31.4660
2364	17	33	58.21	14	8	50.05	263.4926	14.1472
2365	17	34	19.45	-30	22	28.42	263.5810	-30.3746
2366	17	35	34.24	-29	41	49.89	263.8926	-29.6972
2367	17	41	7.59	3	18	9.86	265.2816	3.3027
2368	17	41	29.77	14	43	1.08	265.3740	14.7170
2369	18	7	42.64	12	56	34.88	271.9277	12.9430
2370	18	8	4.09	3	38	23.58	272.0170	3.6399
2371	18	11	14.75	4	28	28.54	272.8115	4.4746
2372	18	11	43.37	11	31	55.25	272.9307	11.5320
2373	18	12	29.27	4	40	21.08	273.1219	4.6725
2374	18	14	9.07	8	1	35.03	273.5378	8.0264
2375	18	22	6.94	6	18	21.99	275.5289	6.3061
2376	18	24	54.33	7	1	35.83	276.2264	7.0266
2377	18	36	30.33	9	18	30.43	279.1264	9.3085

ORION

#	H	M	S	D	M	S	RA DECIMAL	DEC DECIMAL
2378	5	26	45.43	-4	8	10.57	81.6893	-4.1363
2379	5	55	29.74	15	26	34.45	88.8739	15.4429
2380	5	56	28.9	20	2	10.92	89.1204	20.0364
2381	5	56	56.59	15	59	27.48	89.2358	15.9910
2382	5	59	2.64	9	29	2.36	89.7610	9.4840
2383	6	0	23.38	19	58	39.72	90.0974	19.9777
2384	6	2	46.64	7	55	0.78	90.6943	7.9169
2385	6	6	28.66	12	55	25.8	91.6194	12.9238
2386	6	7	8.37	12	24	33.05	91.7849	12.4092
2387	6	7	53.56	0	22	28.75	91.9732	0.3747
2388	6	8	34.66	9	1	12.82	92.1444	9.0202
2389	6	10	17.48	3	56	30.05	92.5728	3.9417
2390	6	11	57.5	12	8	39.22	92.9896	12.1442
2391	6	12	36.57	9	57	40.1	93.1524	9.9611
2392	6	12	57.38	8	6	8.12	93.2391	8.1023
2393	6	13	16.39	3	7	50.45	93.3183	3.1307
2394	6	13	53.89	8	49	7.7	93.4746	8.8188
2395	6	14	34.21	9	40	29.33	93.6425	9.6748
2396	6	15	30.01	11	57	9.74	93.8751	11.9527
2397	6	16	3.23	6	41	22.9	94.0135	6.6897
2398	6	16	53.14	18	28	12.68	94.2214	18.4702
2399	6	18	59.56	8	53	21.2	94.7482	8.8892
2400	6	19	6.55	11	51	29.4	94.7773	11.8582
2401	6	19	28.4	17	34	30.62	94.8683	17.5752

PAVO

#	H	M	S	D	M	S	RA DECIMAL	DEC DECIMAL
2402	18	16	58.05	-59	33	46.05	274.2419	-59.5628
2403	18	20	12.33	-61	44	13.51	275.0514	-61.7371
2404	18	22	57.93	-68	32	12.43	275.7414	-68.5368
2405	19	36	2.73	-69	4	10.94	294.0114	-69.0697
2406	19	36	23.72	-63	31	24.55	294.0988	-63.5235
2407	21	12	4.73	-66	17	59.71	318.0197	-66.2999

PEGASUS

#	H	M	S	D	M	S	RA DECIMAL	DEC DECIMAL
2408	22	33	57.81	30	6	26.63	338.4909	30.1074
2409	22	51	32.49	27	22	47.53	342.8854	27.3799

PERSEUS

#	H	M	S	D	M	S	RA DECIMAL	DEC DECIMAL
2410	1	51	8.09	49	3	28.36	27.7837	49.0579
2411	2	1	31.18	54	59	4.7	30.3799	54.9846
2412	2	1	43.51	55	0	28.44	30.4313	55.0079
2413	2	6	33.69	51	12	19.28	31.6404	51.2054
2414	2	9	1.89	55	14	1.6	32.2579	55.2338
2415	2	9	30.33	57	57	38.27	32.3764	57.9606
2416	2	10	10.66	51	43	0.26	32.5444	51.7167
2417	2	10	13.41	57	11	25	32.5559	57.1903
2418	2	13	55.06	53	53	0.07	33.4794	53.8834
2419	2	14	14.82	57	15	50.74	33.5617	57.2641
2420	2	14	20.29	53	54	20.8	33.5845	53.9058
2421	2	19	48.79	56	23	44.16	34.9533	56.3956
2422	2	21	29.4	53	40	46.98	35.3725	53.6797
2423	2	21	29.44	52	20	37.87	35.3726	52.3439
2424	2	26	14.84	57	17	21.35	36.5618	57.2893
2425	2	32	25.75	55	37	21.28	38.1073	55.6226
2426	2	32	26.47	56	41	41.42	38.1103	56.6948
2427	2	38	38.96	48	40	15.4	39.6623	48.6709
2428	2	39	50.76	50	34	38.2	39.9615	50.5773
2429	2	40	36.95	48	52	2.85	40.1540	48.8675
2430	2	40	44.8	49	27	14.15	40.1866	49.4539
2431	2	41	47.42	40	16	48.34	40.4476	40.2801
2432	2	44	8.34	47	31	20.05	41.0348	47.5222
2433	3	1	14.11	48	51	34.55	45.3088	48.8596
2434	3	11	56.83	50	19	6.16	47.9868	50.3184
2435	3	27	45.85	34	26	35.54	51.9410	34.4432
2436	3	46	32.37	47	58	22	56.6349	47.9728
2437	3	53	2.29	45	57	24.73	58.2596	45.9569
2438	4	6	47.57	43	12	54.9	61.6982	43.2153
2439	4	11	44.48	37	42	6.34	62.9353	37.7018
2440	4	52	29.61	51	22	2.91	73.1234	51.3675

PICTOR

#	H	M	S	D	M	S	RA DECIMAL	DEC DECIMAL
2441	6	50	21.27	-54	59	42.7	102.5886	-54.9952

Nicholson #2441 – 15 x 15 arc minutes

PUPPIS

#	H	M	S	D	M	S	RA DECIMAL	DEC DECIMAL
2442	6	21	45.05	-46	46	14.45	95.4377	-46.7707
2443	6	38	56.67	-45	3	13.2	99.7361	-45.0537
2444	6	53	36.18	-42	12	30.7	103.4008	-42.2085
2445	7	6	37.73	-37	43	11.6	106.6572	-37.7199
2446	7	9	27.44	-43	51	39.43	107.3643	-43.8610
2447	7	11	4.28	-39	23	13.64	107.7679	-39.3871
2448	7	12	26.63	-47	41	47.85	108.1109	-47.6966
2449	7	19	15.37	-45	16	22.28	109.8140	-45.2729
2450	7	20	54.5	-32	17	42.74	110.2271	-32.2952
2451	7	21	10.05	-32	19	10.47	110.2919	-32.3196
2452	7	25	4.6	-32	49	18.94	111.2692	-32.8219
2453	7	25	18.34	-17	58	24.61	111.3264	-17.9735
2454	7	25	20.57	-33	5	41.32	111.3357	-33.0948
2455	7	25	42.9	-17	56	24.63	111.4288	-17.9402
2456	7	26	13.07	-12	20	15.36	111.5545	-12.3376
2457	7	26	51.17	-15	56	24.52	111.7132	-15.9401

2458	7	26	58.37	-13	37	11.54	111.7432	-13.6199
2459	7	27	3.13	-15	54	35.38	111.7631	-15.9098
2460	7	27	33.65	-31	41	30.3	111.8902	-31.6918
2461	7	27	42.3	-18	59	47.3	111.9262	-18.9965
2462	7	27	56.44	-16	23	2.39	111.9852	-16.3840
2463	7	28	14.4	-32	7	14.75	112.0600	-32.1208
2464	7	28	14.88	-18	50	17.53	112.0620	-18.8382
2465	7	28	27.79	-12	47	50.21	112.1158	-12.7973
2466	7	28	35.78	-34	19	52.41	112.1491	-34.3312
2467	7	29	12.95	-31	51	0.24	112.3039	-31.8501
2468	7	32	50.49	-18	24	2.86	113.2104	-18.4008
2469	7	34	1.97	-31	31	14.46	113.5082	-31.5207
2470	7	34	34.34	-21	18	23.36	113.6431	-21.3065
2471	7	35	12.2	-21	55	33.03	113.8008	-21.9258
2472	7	35	32.35	-18	18	19.11	113.8848	-18.3053
2473	7	35	39.21	-33	11	29.15	113.9134	-33.1914
2474	7	35	44.26	-10	1	34.18	113.9344	-10.0262
2475	7	36	2.93	-44	38	59.98	114.0122	-44.6500
2476	7	36	40.95	-22	32	8.13	114.1706	-22.5356
2477	7	36	53.25	-12	48	42.22	114.2219	-12.8117
2478	7	37	11.9	-24	31	40.54	114.2996	-24.5279
2479	7	37	43.7	-12	46	38.17	114.4321	-12.7773
2480	7	37	51.19	-44	0	34.83	114.4633	-44.0097
2481	7	38	24.35	-21	59	9.31	114.6015	-21.9859
2482	7	38	30.44	-18	50	0.99	114.6268	-18.8336
2483	7	38	40.05	-20	49	37.27	114.6669	-20.8270
2484	7	39	0.28	-21	53	34.79	114.7512	-21.8930
2485	7	39	2.16	-20	15	33.1	114.7590	-20.2592
2486	7	39	2.98	-21	50	44.26	114.7624	-21.8456
2487	7	39	14.43	-20	17	8.64	114.8101	-20.2857
2488	7	39	45.61	-26	9	48.58	114.9400	-26.1635
2489	7	40	4.54	-17	53	9.79	115.0189	-17.8861
2490	7	40	13.91	-42	18	17.28	115.0579	-42.3048
2491	7	40	28.29	-23	33	54.79	115.1179	-23.5652
2492	7	40	32.36	-25	29	37.37	115.1348	-25.4937
2493	7	40	35.47	-23	26	40.25	115.1478	-23.4445
2494	7	40	39.05	-25	27	38.98	115.1627	-25.4608
2495	7	40	41.37	-23	28	40.14	115.1724	-23.4778
2496	7	40	41.5	-25	27	13.91	115.1729	-25.4539
2497	7	40	52.21	-25	30	12.78	115.2175	-25.5036
2498	7	40	55.34	-25	27	16.1	115.2306	-25.4545
2499	7	41	3.49	-21	51	46.28	115.2645	-21.8629
2500	7	41	12.19	-19	35	41.06	115.3008	-19.5947

2501	7	41	12.72	-29	35	18.02	115.3030	-29.5883
2502	7	41	16.8	-21	51	44.63	115.3200	-21.8624
2503	7	41	28.47	-10	21	46.1	115.3686	-10.3628
2504	7	42	40.56	-21	58	4.97	115.6690	-21.9680
2505	7	42	44.18	-11	41	8.17	115.6841	-11.6856
2506	7	42	46.24	-20	14	33.3	115.6927	-20.2426
2507	7	42	48.58	-41	19	26.92	115.7024	-41.3241
2508	7	43	4.74	-19	26	6.04	115.7697	-19.4350
2509	7	43	8.2	-33	19	31.84	115.7841	-33.3255
2510	7	43	9.76	-19	24	42.57	115.7907	-19.4118
2511	7	43	17.45	-22	53	9.49	115.8227	-22.8860
2512	7	43	18.09	-19	27	9.51	115.8254	-19.4526
2513	7	43	24.66	-27	25	10.49	115.8528	-27.4196
2514	7	43	27.41	-14	2	48.15	115.8642	-14.0467
2515	7	43	29.73	-27	26	41.55	115.8739	-27.4449
2516	7	43	30.29	-14	4	32.94	115.8762	-14.0758
2517	7	43	32.44	-27	21	54.56	115.8852	-27.3652
2518	7	43	46.82	-12	41	53.21	115.9451	-12.6981
2519	7	43	48.06	-39	56	14.38	115.9503	-39.9373
2520	7	43	52.8	-12	38	27.09	115.9700	-12.6409
2521	7	44	12.32	-21	20	50.59	116.0513	-21.3474
2522	7	44	25.21	-43	12	41.13	116.1050	-43.2114
2523	7	44	30.97	-30	15	58.44	116.1291	-30.2662
2524	7	45	23.32	-29	34	7.02	116.3472	-29.5686
2525	7	45	30.78	-13	23	51.08	116.3783	-13.3975
2526	7	45	45.31	-38	58	12.14	116.4388	-38.9700
2527	7	45	47.98	-14	31	56.64	116.4499	-14.5324
2528	7	46	20.38	-16	42	37.26	116.5849	-16.7103
2529	7	46	31.1	-43	0	2.08	116.6296	-43.0006
2530	7	46	43.32	-43	34	1.91	116.6805	-43.5672
2531	7	46	48.6	-23	16	43.24	116.7025	-23.2787
2532	7	46	49.15	-24	10	17.1	116.7048	-24.1714
2533	7	46	51.12	-29	53	58.03	116.7130	-29.8995
2534	7	46	56.47	-23	0	17.44	116.7353	-23.0048
2535	7	47	5.03	-22	58	55.92	116.7710	-22.9822
2536	7	47	59.89	-25	4	35.21	116.9996	-25.0764
2537	7	49	18.54	-23	47	56.59	117.3273	-23.7991
2538	7	50	4.86	-28	58	54.64	117.5202	-28.9818
2539	7	50	19.27	-15	47	41.32	117.5803	-15.7948
2540	7	50	33.16	-29	16	17.29	117.6382	-29.2715
2541	7	50	40.58	-16	7	1.35	117.6691	-16.1170
2542	7	50	50.52	-18	54	34.27	117.7105	-18.9095
2543	7	50	55	-19	38	37.48	117.7292	-19.6437

2544	7	51	1.91	-33	40	36.55	117.7579	-33.6768
2545	7	51	6.43	-14	9	55	117.7768	-14.1653
2546	7	51	15.74	-43	31	38.84	117.8156	-43.5275
2547	7	51	58.13	-16	45	0.72	117.9922	-16.7502
2548	7	52	8.78	-21	13	30.82	118.0366	-21.2252
2549	7	52	10.04	-24	33	13.43	118.0418	-24.5537
2550	7	52	29.18	-19	40	14.29	118.1216	-19.6706
2551	7	52	37.23	-44	4	27.91	118.1551	-44.0744
2552	7	52	57.21	-25	42	43.73	118.2384	-25.7121
2553	7	53	18.09	-19	29	4.41	118.3254	-19.4846
2554	7	53	24.6	-42	52	21.93	118.3525	-42.8728
2555	7	53	56.8	-22	56	9.13	118.4867	-22.9359
2556	7	55	19.87	-23	1	42.19	118.8328	-23.0284
2557	7	55	33.5	-27	19	9.05	118.8896	-27.3192
2558	7	55	35.22	-22	58	57.66	118.8968	-22.9827
2559	7	55	56.14	-39	29	4.2	118.9839	-39.4845
2560	7	56	3.96	-23	1	41.46	119.0165	-23.0282
2561	7	56	17.85	-44	14	46.61	119.0744	-44.2463
2562	7	56	29.61	-41	59	41.36	119.1234	-41.9948
2563	7	56	30.51	-20	51	23.9	119.1271	-20.8566
2564	7	56	41.13	-12	10	56.53	119.1714	-12.1824
2565	7	58	32.15	-19	29	59.29	119.6340	-19.4998
2566	7	58	53.28	-25	28	34.83	119.7220	-25.4763
2567	7	59	36.76	-21	53	22.04	119.9032	-21.8895
2568	7	59	54.07	-21	3	19.98	119.9753	-21.0556
2569	8	0	11.54	-31	56	22.44	120.0481	-31.9396
2570	8	0	24.7	-25	38	7.97	120.1029	-25.6355
2571	8	0	29.93	-26	37	40.08	120.1247	-26.6278
2572	8	0	56.3	-38	53	32.68	120.2346	-38.8924
2573	8	0	58.88	-31	34	50.96	120.2453	-31.5808
2574	8	1	8.33	-33	37	32.87	120.2847	-33.6258
2575	8	1	18.78	-40	12	41.83	120.3282	-40.2116
2576	8	1	23.49	-33	37	20.29	120.3479	-33.6223
2577	8	1	24.22	-32	53	52.21	120.3509	-32.8978
2578	8	2	35.07	-24	43	21.83	120.6461	-24.7227
2579	8	2	49.54	-17	43	35.56	120.7064	-17.7265
2580	8	3	10.93	-24	36	24.63	120.7955	-24.6068
2581	8	4	23.16	-27	27	23.23	121.0965	-27.4565
2582	8	4	27.36	-36	26	41.81	121.1140	-36.4449
2583	8	4	40.81	-33	2	33.37	121.1701	-33.0426
2584	8	4	43.22	-29	43	11.76	121.1801	-29.7199
2585	8	4	44.07	-33	5	50.04	121.1836	-33.0972
2586	8	5	7.85	-38	22	50.04	121.2827	-38.3806

2587	8	5	15.64	-33	3	6.14	121.3152	-33.0517
2588	8	5	29.78	-18	18	15.85	121.3741	-18.3044
2589	8	5	31.94	-27	9	50.58	121.3831	-27.1640
2590	8	5	34.41	-18	18	26.8	121.3934	-18.3074
2591	8	5	39.09	-27	9	34.04	121.4129	-27.1595
2592	8	5	57.76	-25	17	12.04	121.4907	-25.2867
2593	8	6	11.94	-33	29	59.2	121.5498	-33.4998
2594	8	6	13.94	-24	19	17.04	121.5581	-24.3214
2595	8	6	14.46	-9	42	55.75	121.5603	-9.7155
2596	8	6	39.22	-17	17	50.05	121.6634	-17.2972
2597	8	7	28.84	-21	28	12.47	121.8702	-21.4701
2598	8	7	44.48	-29	24	9.39	121.9353	-29.4026
2599	8	8	2.81	-28	5	6.13	122.0117	-28.0850
2600	8	8	6.8	-20	52	54.27	122.0283	-20.8817
2601	8	8	44.93	-21	28	50	122.1872	-21.4806
2602	8	8	47.66	-42	6	45.11	122.1986	-42.1125
2603	8	9	4.16	-16	22	32.65	122.2673	-16.3757
2604	8	9	5.26	-42	4	28.51	122.2719	-42.0746
2605	8	9	7.91	-31	24	52.74	122.2830	-31.4147
2606	8	9	8.07	-39	34	20.16	122.2836	-39.5723
2607	8	9	15.53	-42	2	8.77	122.3147	-42.0358
2608	8	9	15.87	-27	24	9.76	122.3161	-27.4027
2609	8	9	23.42	-29	48	19.66	122.3476	-29.8055
2610	8	9	29.87	-19	34	44.74	122.3744	-19.5791
2611	8	9	54.26	-21	46	55.6	122.4761	-21.7821
2612	8	10	11.88	-36	7	26.27	122.5495	-36.1240
2613	8	10	21.51	-20	53	49.21	122.5896	-20.8970
2614	8	11	4.26	-19	0	52.83	122.7678	-19.0147
2615	8	11	9.17	-36	16	19.73	122.7882	-36.2721
2616	8	11	15.37	-26	25	28.66	122.8140	-26.4246
2617	8	11	17.15	-29	18	36.28	122.8215	-29.3101
2618	8	11	46.49	-29	37	12.93	122.9437	-29.6203
2619	8	12	42.47	-35	5	51.49	123.1770	-35.0976
2620	8	13	0.39	-23	54	16.29	123.2516	-23.9045
2621	8	14	29.07	-23	31	6.01	123.6211	-23.5183
2622	8	14	46.06	-29	48	26.71	123.6919	-29.8074
2623	8	15	12.34	-21	17	42.69	123.8014	-21.2952
2624	8	15	13.7	-29	37	4.29	123.8071	-29.6179
2625	8	15	15.94	-29	3	33.35	123.8164	-29.0593
2626	8	15	29.6	-29	36	53.04	123.8733	-29.6147
2627	8	15	52.84	-23	8	9.8	123.9702	-23.1361
2628	8	16	25.89	-33	4	56.83	124.1079	-33.0825
2629	8	16	51.47	-39	56	21.72	124.2145	-39.9394

2630	8	18	8.17	-33	57	20.2	124.5341	-33.9556
2631	8	18	22.53	-37	53	10.38	124.5939	-37.8862
2632	8	18	28.77	-26	55	30.72	124.6199	-26.9252
2633	8	19	13.72	-33	24	55.32	124.8072	-33.4154
2634	8	19	55.96	-28	20	18.55	124.9832	-28.3385
2635	8	20	1.16	-25	53	22.99	125.0048	-25.8897
2636	8	20	10.74	-31	11	35.43	125.0448	-31.1932
2637	8	20	18.7	-33	16	35.17	125.0779	-33.2764
2638	8	21	11.55	-34	49	54.26	125.2981	-34.8317
2639	8	21	35.48	-37	27	1.13	125.3978	-37.4503
2640	8	21	43.45	-31	33	56.77	125.4310	-31.5658
2641	8	21	48.49	-27	28	12.73	125.4521	-27.4702
2642	8	21	49.02	-34	28	50.11	125.4542	-34.4806
2643	8	22	5.53	-30	59	56.38	125.5230	-30.9990
2644	8	22	18.73	-34	8	4.2	125.5780	-34.1345
2645	8	22	42.76	-37	52	8.21	125.6782	-37.8689
2646	8	22	52.35	-31	36	28.56	125.7181	-31.6079
2647	8	23	13.11	-31	34	31.63	125.8046	-31.5755
2648	8	23	22.66	-25	2	1.62	125.8444	-25.0338
2649	8	24	26.09	-26	16	40.86	126.1087	-26.2780
2650	8	25	22.69	-37	51	14.57	126.3445	-37.8540
2651	8	25	24.32	-32	24	48.07	126.3513	-32.4134
2652	8	25	49.05	-36	12	29.09	126.4544	-36.2081
2653	8	26	37.22	-32	14	37.87	126.6551	-32.2439
2654	8	26	43.8	-31	20	42.93	126.6825	-31.3453
2655	8	26	44.19	-32	14	39.98	126.6841	-32.2444
2656	8	26	45.76	-31	20	45.97	126.6907	-31.3461
2657	8	27	4.56	-32	31	38.18	126.7690	-32.5273
2658	8	28	2.66	-31	24	6.08	127.0111	-31.4017
2659	8	28	4.62	-33	8	28.25	127.0193	-33.1412
2660	8	28	9.7	-31	23	1.18	127.0404	-31.3837
2661	8	28	28.49	-30	19	18.26	127.1187	-30.3217
2662	8	29	28.97	-32	7	17.58	127.3707	-32.1216
2663	8	30	40.17	-36	41	43.38	127.6674	-36.6954
2664	8	31	59.98	-32	50	3.42	127.9999	-32.8343
2665	8	32	26.84	-26	25	52.87	128.1118	-26.4314

PYXIS

#	H	M	S	D	M	S	RA DECIMAL	DEC DECIMAL
2666	8	26	36.25	-23	3	27.72	126.6510	-23.0577
2667	8	30	1.7	-34	11	10.56	127.5071	-34.1863
2668	8	31	32.29	-35	23	3.7	127.8845	-35.3844
2669	8	34	5.7	-30	18	11.97	128.5238	-30.3033
2670	8	34	13.8	-25	52	6.35	128.5575	-25.8684
2671	8	36	23	-23	58	36.13	129.0958	-23.9767
2672	8	36	40.28	-32	36	4.15	129.1678	-32.6012
2673	8	37	41.74	-32	29	40.3	129.4239	-32.4945
2674	8	38	35.49	-32	2	38.07	129.6479	-32.0439
2675	8	38	50.5	-21	44	10.2	129.7104	-21.7362
2676	8	39	19.72	-36	1	4.53	129.8322	-36.0179
2677	8	40	3.54	-26	29	35.38	130.0148	-26.4932
2678	8	40	29.35	-29	56	51.16	130.1223	-29.9475
2679	8	40	57.67	-31	51	25.08	130.2403	-31.8570
2680	8	44	6.68	-30	42	29.91	131.0278	-30.7083
2681	8	46	2.59	-22	43	25.97	131.5108	-22.7239
2682	8	50	52.6	-37	16	19.39	132.7192	-37.2721
2683	8	56	3.2	-37	26	20.78	134.0133	-37.4391
2684	9	1	26.74	-30	42	59.82	135.3614	-30.7166
2685	9	13	28.47	-35	54	55.1	138.3686	-35.9153
2686	9	13	51.97	-37	19	32.54	138.4666	-37.3257
2687	9	14	17.3	-37	18	31.27	138.5721	-37.3087
2688	9	14	31.54	-29	44	25.68	138.6314	-29.7405
2689	9	15	31.99	-34	19	3.27	138.8833	-34.3176
2690	9	16	42.82	-30	11	12.88	139.1784	-30.1869

SCULPTOR

#	H	M	S	D	M	S	RA DECIMAL	DEC DECIMAL
2691	23	59	59.83	-27	43	8.46	359.9993	-27.7190

SCORPIUS

#	H	M	S	D	M	S	RA DECIMAL	DEC DECIMAL
2692	15	54	27.2	-27	43	59.63	238.6133	-27.7332
2693	15	59	33.28	-44	52	26.82	239.8887	-44.8741
2694	16	22	38.18	-31	59	2.55	245.6591	-31.9840
2695	16	24	6.63	-33	9	30.34	246.0276	-33.1584
2696	16	32	34.67	-45	25	44.91	248.1445	-45.4291
2697	16	36	34.9	-31	29	24.73	249.1454	-31.4902
2698	16	36	39.53	-31	28	5.05	249.1647	-31.4681
2699	16	39	39.57	-34	11	35.92	249.9149	-34.1933
2700	16	46	48.92	-39	15	27.95	251.7038	-39.2578
2701	16	46	59.62	-41	12	25.32	251.7484	-41.2070
2702	16	56	16.46	-37	15	44.55	254.0686	-37.2624
2703	16	58	27.1	-38	25	45.35	254.6129	-38.4293
2704	17	1	19.38	-38	10	57.22	255.3308	-38.1826
2705	17	3	32.86	-44	17	22.19	255.8869	-44.2895
2706	17	5	6.69	-43	36	42.59	256.2779	-43.6118
2707	17	5	32.64	-45	3	22.02	256.3860	-45.0561
2708	17	5	44.6	-44	32	40.48	256.4358	-44.5446
2709	17	6	19.72	-42	52	15.74	256.5822	-42.8710
2710	17	8	4.54	-38	30	13.18	257.0189	-38.5037
2711	17	11	23.32	-33	2	38.93	257.8472	-33.0441
2712	17	12	24.65	-37	42	28.66	258.1027	-37.7080
2713	17	12	54.34	-32	20	52.31	258.2264	-32.3479
2714	17	14	42.99	-33	16	23.83	258.6791	-33.2733
2715	17	15	53.08	-32	43	4.67	258.9712	-32.7180
2716	17	17	46.35	-33	11	50.43	259.4431	-33.1973
2717	17	18	44.18	-41	5	57.91	259.6841	-41.0994
2718	17	19	18.47	-33	4	0.93	259.8269	-33.0669
2719	17	19	37.39	-41	52	15.28	259.9058	-41.8709
2720	17	20	35.75	-29	20	48.75	260.1490	-29.3469
2721	17	20	55.73	-39	59	36.65	260.2322	-39.9935
2722	17	21	27.44	-41	52	6.81	260.3643	-41.8686
2723	17	21	34.67	-41	14	12.72	260.3945	-41.2369
2724	17	22	48.58	-41	58	26.51	260.7024	-41.9740
2725	17	23	29.85	-39	34	44.03	260.8744	-39.5789
2726	17	24	28.93	-45	9	34.16	261.1206	-45.1595
2727	17	24	46.33	-45	10	36.51	261.1931	-45.1768
2728	17	24	48.75	-41	39	10.21	261.2031	-41.6528
2729	17	25	8.03	-39	34	49.81	261.2835	-39.5805
2730	17	25	17.15	-44	22	15.52	261.3215	-44.3710

2731	17	25	22.48	-45	14	44.9	261.3437	-45.2458
2732	17	25	22.7	-44	22	1.14	261.3446	-44.3670
2733	17	25	34.26	-45	15	26.02	261.3928	-45.2572
2734	17	26	8.34	-44	7	27.42	261.5347	-44.1243
2735	17	26	42.76	-44	39	32.6	261.6782	-44.6591
2736	17	27	0.71	-43	5	48.71	261.7529	-43.0969
2737	17	27	5.03	-40	8	15.53	261.7710	-40.1376
2738	17	27	6.14	-44	11	48.56	261.7756	-44.1968
2739	17	27	27.06	-39	3	36.07	261.8627	-39.0600
2740	17	27	32.9	-45	38	14.35	261.8871	-45.6373
2741	17	27	35.21	-38	25	24.92	261.8967	-38.4236
2742	17	27	47.96	-44	46	13.53	261.9498	-44.7704
2743	17	28	14.22	-43	44	43.59	262.0592	-43.7454
2744	17	28	25.6	-45	54	22.23	262.1067	-45.9062
2745	17	28	27.26	-36	54	19.79	262.1136	-36.9055
2746	17	28	31.18	-44	39	0.09	262.1299	-44.6500
2747	17	29	46.77	-44	25	5.96	262.4449	-44.4183
2748	17	30	29.7	-31	6	58.61	262.6238	-31.1163
2749	17	31	3.72	-42	8	11.7	262.7655	-42.1366
2750	17	31	26.5	-45	4	34.82	262.8604	-45.0763
2751	17	31	50.61	-38	44	27.84	262.9609	-38.7411
2752	17	32	15.35	-44	22	39.29	263.0639	-44.3776
2753	17	32	36.82	-44	4	25.64	263.1534	-44.0738
2754	17	32	51.92	-43	19	59.47	263.2163	-43.3332
2755	17	33	31.88	-39	25	30.13	263.3828	-39.4250
2756	17	33	43.27	-46	9	15	263.4303	-46.1542
2757	17	34	46.06	-43	4	19.19	263.6919	-43.0720
2758	17	35	12.54	-41	7	1.96	263.8022	-41.1172
2759	17	36	39.77	-43	1	33.5	264.1657	-43.0260
2760	17	36	58.19	-42	59	5.34	264.2425	-42.9848
2761	17	37	46.92	-41	16	16.86	264.4455	-41.2713
2762	17	38	27.49	-40	13	17.46	264.6146	-40.2215
2763	17	38	27.79	-42	46	54.85	264.6158	-42.7819
2764	17	39	49.75	-45	30	27.4	264.9573	-45.5076
2765	17	41	7.88	-39	25	44.06	265.2828	-39.4289
2766	17	41	8.93	-43	47	30.84	265.2872	-43.7919
2767	17	41	12.99	-41	40	45.25	265.3041	-41.6792
2768	17	41	59.07	-33	29	29.76	265.4961	-33.4916
2769	17	41	59.72	-40	40	27.77	265.4988	-40.6744
2770	17	42	26.1	-41	13	18.66	265.6088	-41.2219
2771	17	42	34.76	-41	15	29.04	265.6449	-41.2581
2772	17	44	46.4	-42	16	33.31	266.1934	-42.2759
2773	17	45	20.31	-41	24	33.58	266.3346	-41.4093

2774	17	45	43.85	-43	38	10.67	266.4327	-43.6363
2775	17	46	58.21	-43	30	46.71	266.7425	-43.5130
2776	17	48	8.51	-40	54	29.48	267.0355	-40.9082
2777	17	48	10.32	-44	4	13.76	267.0430	-44.0705
2778	17	48	21.68	-40	57	38.47	267.0903	-40.9607
2779	17	48	59.31	-45	5	23.44	267.2471	-45.0898
2780	17	49	16.12	-41	23	33.89	267.3171	-41.3927
2781	17	49	23.2	-40	51	3.86	267.3467	-40.8511
2782	17	49	26.27	-43	5	50.13	267.3595	-43.0973
2783	17	49	33.13	-32	54	36.58	267.3881	-32.9102
2784	17	49	56.9	-38	56	24.32	267.4871	-38.9401
2785	17	50	0.53	-36	20	36.39	267.5022	-36.3434
2786	17	50	21.75	-44	53	44.54	267.5906	-44.8957
2787	17	50	22.25	-41	30	47.05	267.5927	-41.5131
2788	17	50	57.22	-33	45	45.49	267.7384	-33.7626
2789	17	50	59.79	-39	10	7.94	267.7491	-39.1689
2790	17	51	15.77	-37	48	34.92	267.8157	-37.8097
2791	17	51	44.3	-42	57	56.36	267.9346	-42.9657
2792	17	52	45.19	-38	55	35.57	268.1883	-38.9265
2793	17	53	22.77	-40	13	57.1	268.3449	-40.2325
2794	17	53	59.38	-42	21	33.91	268.4974	-42.3594
2795	17	53	59.44	-33	16	42.04	268.4977	-33.2783
2796	17	54	2.12	-42	20	5.54	268.5088	-42.3349
2797	17	54	5.04	-42	54	0.49	268.5210	-42.9001
2798	17	54	7.21	-39	16	25.21	268.5301	-39.2737
2799	17	54	10.36	-38	27	57.95	268.5432	-38.4661
2800	17	54	29.2	-35	56	8.68	268.6217	-35.9357
2801	17	54	31.75	-41	52	34.81	268.6323	-41.8763
2802	17	54	40.2	-31	3	14.47	268.6675	-31.0540
2803	17	54	49.73	-33	54	52.97	268.7072	-33.9147
2804	17	54	51.71	-40	17	19.43	268.7155	-40.2887
2805	17	54	59.22	-39	22	44.19	268.7468	-39.3789
2806	17	55	2.73	-40	22	20.11	268.7614	-40.3723
2807	17	55	54.97	-31	24	56.18	268.9790	-31.4156
2808	17	56	0.22	-38	9	11.86	269.0009	-38.1533
2809	17	56	13.29	-41	1	52.91	269.0554	-41.0314
2810	17	56	23.63	-29	48	19.37	269.0985	-29.8054
2811	17	56	27.99	-29	49	19.29	269.1166	-29.8220
2812	17	56	28.14	-38	31	15.25	269.1172	-38.5209
2813	17	56	53.24	-34	34	43.74	269.2218	-34.5788
2814	17	56	57.9	-42	35	49.71	269.2412	-42.5971
2815	17	57	6.33	-38	27	30.23	269.2764	-38.4584
2816	17	57	10.6	-30	18	48.44	269.2942	-30.3135

2817	17	57	10.7	-39	56	6.43	269.2946	-39.9351
2818	17	57	22.06	-39	57	16.53	269.3419	-39.9546
2819	17	57	27.76	-30	34	9.2	269.3656	-30.5692
2820	17	57	31.15	-40	9	50.78	269.3798	-40.1641
2821	17	57	42.9	-38	57	33.47	269.4288	-38.9593
2822	17	57	53.38	-34	10	52.13	269.4724	-34.1811
2823	17	58	16.41	-46	14	15.52	269.5684	-46.2376
2824	17	58	31.21	-38	8	27.62	269.6300	-38.1410
2825	17	58	46.29	-30	23	26.2	269.6929	-30.3906
2826	17	58	49.28	-31	28	19.08	269.7053	-31.4720
2827	17	58	54.05	-38	1	56.43	269.7252	-38.0323
2828	17	59	6.85	-45	42	14.57	269.7785	-45.7040
2829	17	59	14.09	-35	17	57.72	269.8087	-35.2994
2830	17	59	28.47	-34	31	12.99	269.8686	-34.5203
2831	17	59	36.34	-45	19	43.61	269.9014	-45.3288
2832	17	59	53.97	-45	17	20.79	269.9749	-45.2891
2833	17	59	56.94	-33	52	10.13	269.9873	-33.8695
2834	18	0	3.53	-33	58	10.44	270.0147	-33.9696
2835	18	0	34.22	-34	49	56.61	270.1426	-34.8324
2836	18	0	58.36	-34	20	56.1	270.2432	-34.3489
2837	18	2	24.13	-45	44	16.64	270.6005	-45.7380

Nicholson #2835 – 15 x 15 arc minutes

SAGITTA

#	H	M	S	D	M	S	RA DECIMAL	DEC DECIMAL
2838	18	22	30.3	-13	6	24.8	275.6264	-13.1069
2839	18	26	30.9	-14	9	6.15	276.6287	-14.1517
2840	18	28	9.32	-15	2	16.1	277.0388	-15.0378
2841	18	33	21	-14	15	19.4	278.3374	-14.2554
2842	18	35	59.6	-13	3	10.3	278.9982	-13.0529
2843	18	42	51.7	-7	17	5.06	280.7154	-7.2847
2844	18	43	8.99	-6	50	57.3	280.7875	-6.8493
2845	18	43	11.4	-6	48	19.3	280.7974	-6.8054
2846	18	43	37.2	-6	53	30.7	280.9050	-6.8919
2847	18	44	42.2	-7	35	28.6	281.1757	-7.5913
2848	18	44	49.7	-6	20	26	281.2069	-6.3405
2849	18	45	7.58	-7	10	59.5	281.2816	-7.1832
2850	18	45	26.3	-6	36	20.6	281.3598	-6.6057
2851	18	45	56.7	-6	50	22.4	281.4860	-6.8396
2852	18	46	19.8	-7	15	47	281.5823	-7.2631
2853	18	47	46.2	-6	34	30.4	281.9425	-6.5751
2854	18	49	14.5	-7	34	13.1	282.3104	-7.5703
2855	18	49	23.9	-6	53	27.6	282.3496	-6.8910
2856	18	49	50.2	-11	28	31.5	282.4590	-11.4754
2857	18	50	25.5	-6	47	33	282.6061	-6.7925
2858	18	51	15.6	-7	48	20.9	282.8150	-7.8058
2859	18	52	22.7	-7	38	44.2	283.0946	-7.6456
2860	18	53	17.5	-7	48	50.7	283.3230	-7.8141
2861	18	54	5.99	-4	40	47.6	283.5250	-4.6799
2862	18	55	60	-6	45	33.1	283.9999	-6.7592

SERPENS

#	H	M	S	D	M	S	RA DECIMAL	DEC DECIMAL
2864	18	13	29.01	-14	30	59.87	273.3709	-14.5166
2865	18	15	35.84	-15	48	54.44	273.8993	-15.8151
2866	18	19	59.29	-12	28	7.13	274.9971	-12.4686
2867	18	37	30.7	8	27	55.02	279.3779	8.4653
2868	18	39	25.69	7	14	32.07	279.8570	7.2422

SAGITTA

#	H	M	S	D	M	S	RA DECIMAL	DEC DECIMAL
2869	19	42	50.19	15	22	52.36	295.7091	15.3812
2870	19	43	58.65	14	55	15.89	295.9944	14.9211
2871	19	45	4.68	16	39	40.2	296.2695	16.6612
2872	19	47	20.1	17	20	21.49	296.8338	17.3393
2873	19	47	53.42	15	54	16.08	296.9726	15.9045
2874	19	48	28.06	18	39	18.43	297.1169	18.6551
2875	19	49	39.96	18	44	36.36	297.4165	18.7434
2876	19	49	47.94	18	44	59.97	297.4497	18.7500
2877	19	54	12.99	15	6	4.83	298.5541	15.1013
2878	19	54	16.71	15	5	50.38	298.5696	15.0973
2879	19	57	8.92	18	59	35.76	299.2872	18.9933
2880	19	57	32.81	16	31	40.2	299.3867	16.5278
2881	19	59	44.66	19	17	56.42	299.9361	19.2990
2882	20	0	10.25	17	38	28.34	300.0427	17.6412
2883	20	0	16.34	15	58	32.94	300.0681	15.9758
2884	20	0	53.87	20	11	18.54	300.2245	20.1885
2885	20	2	33.73	18	34	43.07	300.6405	18.5786
2886	20	4	5.24	21	22	53.12	301.0218	21.3814
2887	20	4	13.01	17	2	43.89	301.0542	17.0455
2888	20	5	50.25	19	18	51.45	301.4594	19.3143
2889	20	6	8.33	15	12	20.51	301.5347	15.2057
2890	20	6	26.3	22	12	47.48	301.6096	22.2132
2891	20	7	11.62	17	34	4.44	301.7984	17.5679
2892	20	7	20.11	15	14	32.16	301.8338	15.2423
2893	20	9	3.04	21	38	52.35	302.2627	21.6479

2894	20	10	12.89	19	15	42.54	302.5537	19.2618
2895	20	10	42.49	22	4	51.76	302.6770	22.0810
2896	20	12	8.69	21	20	3.63	303.0362	21.3343
2897	20	12	19.14	17	11	34.8	303.0797	17.1930
2898	20	13	23.25	17	33	10.5	303.3469	17.5529
2899	20	13	24.01	17	37	58.11	303.3500	17.6328
2900	20	13	53.32	19	3	10.71	303.4722	19.0530
2901	20	14	17.62	17	47	17.53	303.5734	17.7882
2902	20	14	31.82	17	37	16.89	303.6326	17.6214
2903	20	15	19.12	20	59	50.83	303.8297	20.9975
2904	20	16	36.86	14	50	4.08	304.1536	14.8345
2905	20	21	30.76	18	23	29.69	305.3781	18.3916
2906	20	28	4.1	16	25	10.55	307.0171	16.4196
2907	20	28	28.27	14	23	15.29	307.1178	14.3876

SAGITTARIUS

#	H	M	S	D	M	S	RA DECIMAL	DEC DECIMAL
2908	17	55	6.88	-29	37	48.77	268.7787	-29.6302
2909	17	55	16.16	-26	4	58.94	268.8173	-26.0830
2910	17	56	12.77	-28	41	54.83	269.0532	-28.6986
2911	17	56	18.95	-29	37	51.23	269.0790	-29.6309
2912	17	56	19.68	-28	25	52.52	269.0820	-28.4313
2913	17	56	25.75	-28	30	30.08	269.1073	-28.5084
2914	17	56	37.14	-29	38	14.32	269.1547	-29.6373
2915	17	57	17.76	-26	37	38.41	269.3240	-26.6273
2916	17	57	45.64	-26	53	22.5	269.4402	-26.8896
2917	17	59	8.03	-15	16	5.84	269.7835	-15.2683
2918	17	59	10.36	-35	58	51.04	269.7932	-35.9808
2919	17	59	14.08	-36	56	21.58	269.8087	-36.9393
2920	17	59	16	-36	0	26.51	269.8167	-36.0074
2921	17	59	26.11	-37	5	37.32	269.8588	-37.0937
2922	17	59	36.34	-30	40	16.67	269.9014	-30.6713
2923	17	59	56.96	-30	2	43.66	269.9873	-30.0455
2924	18	0	9.65	-37	18	42.37	270.0402	-37.3118
2925	18	0	12.39	-27	36	34.59	270.0516	-27.6096
2926	18	0	15.07	-37	24	29.44	270.0628	-37.4082
2927	18	0	19.14	-30	6	45.61	270.0798	-30.1127
2928	18	0	21.59	-37	28	59.02	270.0899	-37.4831
2929	18	0	25.4	-30	3	45.44	270.1059	-30.0626
2930	18	0	27.3	-22	21	4.79	270.1138	-22.3513
2931	18	0	29.2	-27	37	54.83	270.1217	-27.6319
2932	18	0	31.01	-30	9	27.47	270.1292	-30.1576
2933	18	0	33.48	-36	53	19.45	270.1395	-36.8887
2934	18	0	49.17	-36	55	51.96	270.2049	-36.9311
2935	18	1	8.73	-29	5	54.92	270.2864	-29.0986
2936	18	1	12.38	-36	11	59.06	270.3016	-36.1997
2937	18	1	18.68	-31	57	44.35	270.3278	-31.9623
2938	18	1	19.27	-36	9	37.03	270.3303	-36.1603
2939	18	1	22.04	-31	54	29.21	270.3418	-31.9081
2940	18	1	27.34	-26	18	7.3	270.3639	-26.3020
2941	18	1	31.3	-29	2	25.1	270.3804	-29.0403
2942	18	1	33.46	-26	19	22.03	270.3894	-26.3228
2943	18	1	36.5	-37	7	27.85	270.4021	-37.1244
2944	18	1	47.09	-36	27	34.08	270.4462	-36.4595
2945	18	1	53.09	-37	10	24.02	270.4712	-37.1733
2946	18	2	9.57	-36	35	54.84	270.5399	-36.5986

2947	18	2	18.3	-38	40	28.45	270.5763	-38.6746
2948	18	2	22.4	-26	1	4.94	270.5933	-26.0180
2949	18	2	26.37	-25	55	37.33	270.6099	-25.9270
2950	18	2	35.33	-29	17	21.26	270.6472	-29.2892
2951	18	2	40.08	-39	50	35.01	270.6670	-39.8431
2952	18	2	41.52	-29	17	19.14	270.6730	-29.2887
2953	18	2	42.19	-39	45	35.9	270.6758	-39.7600
2954	18	2	46.5	-30	58	11.28	270.6938	-30.9698
2955	18	2	58.54	-28	27	34.56	270.7439	-28.4596
2956	18	3	7.58	-28	45	52.87	270.7816	-28.7647
2957	18	3	9.84	-37	45	40.16	270.7910	-37.7612
2958	18	3	12.24	-28	50	31.81	270.8010	-28.8422
2959	18	3	16.84	-37	43	32.77	270.8202	-37.7258
2960	18	3	21.21	-28	53	34.1	270.8384	-28.8928
2961	18	3	31.65	-37	44	42.88	270.8819	-37.7452
2962	18	3	35.99	-29	6	15.95	270.9000	-29.1044
2963	18	3	39.1	-37	43	47.67	270.9129	-37.7299
2964	18	3	43.43	-25	27	24.25	270.9310	-25.4567
2965	18	3	49.57	-28	44	15.51	270.9566	-28.7376
2966	18	3	59.09	-25	28	38.38	270.9962	-25.4773
2967	18	3	59.16	-28	48	14.18	270.9965	-28.8039
2968	18	3	59.19	-25	28	33.79	270.9966	-25.4761
2969	18	4	2.45	-28	49	1.6	271.0102	-28.8171
2970	18	4	24.26	-29	18	29.84	271.1011	-29.3083
2971	18	4	27.16	-35	19	9.8	271.1132	-35.3194
2972	18	4	32.43	-29	19	5.13	271.1351	-29.3181
2973	18	4	33.31	-37	16	19.71	271.1388	-37.2721
2974	18	4	39.58	-26	1	7.27	271.1649	-26.0187
2975	18	4	44.54	-37	17	26.4	271.1856	-37.2907
2976	18	4	50.47	-34	43	14.21	271.2103	-34.7206
2977	18	4	55.78	-32	15	33.87	271.2324	-32.2594
2978	18	5	6.68	-39	23	16.92	271.2778	-39.3880
2979	18	5	13.08	-25	55	1	271.3045	-25.9169
2980	18	5	27.97	-26	31	20.04	271.3666	-26.5222
2981	18	5	30.76	-29	9	5.16	271.3782	-29.1514
2982	18	5	31.91	-26	30	26.91	271.3829	-26.5075
2983	18	5	35.94	-29	5	55.02	271.3997	-29.0986
2984	18	5	38.25	-26	8	5.97	271.4094	-26.1350
2985	18	5	44	-28	48	51.73	271.4333	-28.8144
2986	18	5	44.71	-34	41	39.57	271.4363	-34.6943
2987	18	5	44.86	-28	46	51.86	271.4369	-28.7811
2988	18	5	47.47	-33	22	32.69	271.4478	-33.3757
2989	18	5	48.09	-29	34	56.37	271.4504	-29.5823

2990	18	5	51.9	-34	37	51.84	271.4663	-34.6311
2991	18	5	55.55	-29	34	49.47	271.4815	-29.5804
2992	18	5	56.45	-34	38	29.91	271.4852	-34.6416
2993	18	6	8.77	-31	33	13.1	271.5366	-31.5536
2994	18	6	20.66	-26	42	47.98	271.5861	-26.7133
2995	18	6	33.23	-29	21	32.2	271.6385	-29.3589
2996	18	6	33.65	-30	41	6.28	271.6402	-30.6851
2997	18	6	57.14	-31	46	21.03	271.7381	-31.7725
2998	18	7	3.85	-34	7	11.28	271.7660	-34.1198
2999	18	7	12.68	-32	46	25.44	271.8028	-32.7737
3000	18	7	25.71	-33	39	6.32	271.8571	-33.6518
3001	18	7	38.49	-36	55	1.19	271.9104	-36.9170
3002	18	7	38.64	-33	41	9	271.9110	-33.6858
3003	18	8	2.86	-27	3	59.54	272.0119	-27.0665
3004	18	8	4.05	-35	20	31.18	272.0169	-35.3420
3005	18	8	4.37	-39	5	48.86	272.0182	-39.0969
3006	18	8	6.42	-27	1	50.59	272.0267	-27.0307
3007	18	8	20.44	-35	23	18.92	272.0852	-35.3886
3008	18	8	23.98	-19	8	37.85	272.0999	-19.1438
3009	18	8	26.26	-39	54	45.95	272.1094	-39.9128
3010	18	8	29.49	-35	26	42.61	272.1229	-35.4452
3011	18	8	47.18	-34	56	11.53	272.1966	-34.9365
3012	18	8	52.64	-38	23	6.91	272.2193	-38.3853
3013	18	8	55.66	-34	56	55.32	272.2319	-34.9487
3014	18	9	2.97	-38	19	35.06	272.2624	-38.3264
3015	18	9	28.78	-33	58	17.03	272.3699	-33.9714
3016	18	9	32.48	-28	59	32.18	272.3853	-28.9923
3017	18	9	33.09	-35	7	16.24	272.3879	-35.1212
3018	18	9	35.52	-28	39	54.49	272.3980	-28.6651
3019	18	9	36.23	-35	5	8.84	272.4009	-35.0858
3020	18	9	36.56	-28	39	20.88	272.4023	-28.6558
3021	18	9	38.2	-35	7	24.21	272.4092	-35.1234
3022	18	9	53.83	-28	38	23.03	272.4743	-28.6397
3023	18	9	56.15	-35	32	18.59	272.4840	-35.5385
3024	18	9	58.72	-28	39	8.98	272.4947	-28.6525
3025	18	10	34.91	-30	46	47.36	272.6455	-30.7798
3026	18	10	41.84	-20	19	58.66	272.6743	-20.3330
3027	18	10	55.95	-35	21	26.7	272.7331	-35.3574
3028	18	11	14.05	-26	29	41.11	272.8086	-26.4948
3029	18	11	43.26	-16	4	45.68	272.9302	-16.0794
3030	18	11	47.22	-18	50	42.85	272.9467	-18.8452
3031	18	11	51.11	-30	5	22.61	272.9630	-30.0896
3032	18	11	52.15	-16	4	46.66	272.9673	-16.0796

3033	18	11	52.27	-34	25	44.56	272.9678	-34.4290
3034	18	12	1.09	-34	38	21.26	273.0045	-34.6392
3035	18	12	4.4	-37	47	18.98	273.0183	-37.7886
3036	18	12	21.56	-17	41	3.88	273.0898	-17.6844
3037	18	12	23.17	-37	54	44.8	273.0966	-37.9124
3038	18	12	23.43	-35	34	0.23	273.0976	-35.5667
3039	18	12	28.62	-18	38	51.04	273.1193	-18.6475
3040	18	12	50.65	-37	11	5.5	273.2111	-37.1849
3041	18	12	56.99	-30	25	37.28	273.2375	-30.4270
3042	18	13	3.55	-37	12	50.66	273.2648	-37.2141
3043	18	13	13.2	-18	19	11.66	273.3050	-18.3199
3044	18	13	16.49	-39	15	15.92	273.3187	-39.2544
3045	18	13	24.88	-18	23	52.63	273.3537	-18.3980
3046	18	13	37.4	-34	30	27.42	273.4059	-34.5076
3047	18	13	51.22	-18	28	13.25	273.4634	-18.4703
3048	18	14	0.65	-27	52	28.51	273.5027	-27.8746
3049	18	14	8.2	-30	53	21.55	273.5342	-30.8893
3050	18	14	10.29	-28	41	47.42	273.5429	-28.6965
3051	18	14	15.91	-35	35	42.51	273.5663	-35.5951
3052	18	14	21.98	-27	30	36.81	273.5916	-27.5102
3053	18	14	46.3	-28	10	39.69	273.6929	-28.1777
3054	18	14	55.63	-35	59	46.01	273.7318	-35.9961
3055	18	14	58.88	-28	11	46.13	273.7454	-28.1961
3056	18	15	8.25	-33	28	32.97	273.7844	-33.4758
3057	18	15	22.67	-29	46	6.83	273.8445	-29.7686
3058	18	15	23.38	-27	31	44.06	273.8474	-27.5289
3059	18	15	27.32	-34	35	40.31	273.8638	-34.5945
3060	18	15	32.16	-15	55	24.52	273.8840	-15.9235
3061	18	15	32.53	-34	32	30.66	273.8855	-34.5419
3062	18	15	36.07	-24	8	12.13	273.9003	-24.1367
3063	18	15	38.81	-27	50	34.9	273.9117	-27.8430
3064	18	15	40.74	-15	53	37.43	273.9198	-15.8937
3065	18	15	42.77	-20	59	29.69	273.9282	-20.9916
3066	18	15	43.57	-27	46	56.18	273.9315	-27.7823
3067	18	15	44.42	-37	42	8.96	273.9351	-37.7025
3068	18	15	45.68	-27	51	15.64	273.9403	-27.8543
3069	18	15	50.52	-23	23	37.75	273.9605	-23.3938
3070	18	16	11.16	-37	40	5.21	274.0465	-37.6681
3071	18	16	17.32	-31	44	35.73	274.0722	-31.7433
3072	18	16	24.38	-30	41	39.47	274.1016	-30.6943
3073	18	16	32.6	-29	42	4.94	274.1358	-29.7014
3074	18	16	38.9	-27	27	57.96	274.1621	-27.4661
3075	18	16	47.69	-29	42	48.05	274.1987	-29.7133

3076	18	16	50.45	-27	23	49.46	274.2102	-27.3971
3077	18	17	2.48	-20	55	23.53	274.2604	-20.9232
3078	18	17	9.68	-28	43	0.19	274.2903	-28.7167
3079	18	17	19.91	-27	35	58.44	274.3330	-27.5996
3080	18	17	30.66	-23	22	4.26	274.3778	-23.3679
3081	18	17	41.23	-28	48	46.25	274.4218	-28.8128
3082	18	17	47.41	-27	49	26.87	274.4475	-27.8241
3083	18	17	53.57	-39	36	6.04	274.4732	-39.6017
3084	18	18	0.11	-17	53	42.19	274.5004	-17.8951
3085	18	18	32.2	-27	55	31.71	274.6342	-27.9255
3086	18	19	14.59	-32	59	60	274.8108	-33.0000
3087	18	19	37.68	-17	58	5.24	274.9070	-17.9681
3088	18	19	55.04	-22	3	55.94	274.9793	-22.0655
3089	18	20	46.3	-19	33	37.25	275.1929	-19.5603
3090	18	20	50.84	-34	23	6.25	275.2118	-34.3851
3091	18	21	3.46	-39	9	18.93	275.2644	-39.1553
3092	18	21	16.64	-35	34	27.44	275.3193	-35.5743
3093	18	21	17.33	-37	14	28.66	275.3222	-37.2413
3094	18	21	25.08	-17	36	34.28	275.3545	-17.6095
3095	18	21	28.09	-37	14	53.15	275.3671	-37.2481
3096	18	21	29.43	-17	40	25.8	275.3726	-17.6738
3097	18	22	32.21	-21	57	44.85	275.6342	-21.9625
3098	18	22	43.29	-34	1	57.33	275.6804	-34.0326
3099	18	22	49.76	-17	42	49.46	275.7074	-17.7137
3100	18	22	54.1	-17	26	1.07	275.7254	-17.4336
3101	18	22	55.57	-34	0	0.91	275.7315	-34.0003
3102	18	23	1.21	-17	10	56.84	275.7551	-17.1825
3103	18	23	6.09	-23	34	49.88	275.7754	-23.5805
3104	18	23	31.08	-36	20	28.57	275.8795	-36.3413
3105	18	23	32.45	-35	0	47.53	275.8852	-35.0132
3106	18	23	36.02	-17	9	55.97	275.9001	-17.1655
3107	18	23	46.52	-15	46	3.47	275.9438	-15.7676
3108	18	23	56.86	-27	22	7.48	275.9869	-27.3687
3109	18	24	53.18	-35	8	53.38	276.2216	-35.1482
3110	18	25	1.47	-38	32	48.95	276.2561	-38.5469
3111	18	25	1.59	-28	28	49.38	276.2566	-28.4804
3112	18	25	28.43	-27	51	43.17	276.3684	-27.8620
3113	18	26	24.65	-29	20	52.88	276.6027	-29.3480
3114	18	26	40.62	-26	48	15.65	276.6693	-26.8043
3115	18	26	53.2	-21	56	18.46	276.7217	-21.9385
3116	18	27	37.45	-35	48	6.62	276.9060	-35.8018
3117	18	27	52.2	-32	8	31.68	276.9675	-32.1421
3118	18	27	52.22	-19	10	27.81	276.9676	-19.1744

3119	18	28	12.27	-36	15	15.62	277.0511	-36.2543
3120	18	29	2.06	-19	10	59.52	277.2586	-19.1832
3121	18	29	48.19	-18	36	1.78	277.4508	-18.6005
3122	18	30	34.3	-16	39	45.4	277.6429	-16.6626
3123	18	30	38.98	-36	47	21.56	277.6624	-36.7893
3124	18	30	40.29	-16	38	46.61	277.6679	-16.6463
3125	18	33	36.88	-38	10	27.26	278.4037	-38.1742
3126	18	33	47.35	-20	21	33.65	278.4473	-20.3593
3127	18	34	6.02	-15	25	30.61	278.5251	-15.4252
3128	18	35	32.82	-18	23	3.02	278.8868	-18.3842
3129	18	36	0.42	-30	23	6.02	279.0018	-30.3850
3130	18	37	4.45	-22	17	22.19	279.2685	-22.2895
3131	18	37	9.06	-16	15	56.35	279.2878	-16.2657
3132	18	38	6.16	-17	16	47.56	279.5256	-17.2799
3133	18	38	57.91	-32	55	43.86	279.7413	-32.9289
3134	18	40	7.34	-38	2	12.64	280.0306	-38.0368
3135	18	42	43.33	-27	33	43.05	280.6806	-27.5620
3136	18	42	54.53	-15	9	13.94	280.7272	-15.1539
3137	18	43	3.59	-18	7	24.08	280.7650	-18.1234
3138	18	43	9.6	-18	9	34.77	280.7900	-18.1597
3139	18	43	11.92	-18	5	40.81	280.7997	-18.0947
3140	18	44	1.37	-15	45	26.35	281.0057	-15.7573
3141	18	45	26.19	-26	57	30.02	281.3591	-26.9583
3142	18	45	57.98	-28	57	5.83	281.4916	-28.9516
3143	18	46	30.49	-29	25	4.68	281.6270	-29.4180
3144	18	48	1.68	-18	17	26.94	282.0070	-18.2908
3145	18	48	34.83	-19	37	42.28	282.1451	-19.6284
3146	18	49	34.17	-17	48	13.55	282.3924	-17.8038
3147	18	49	34.32	-12	45	26.57	282.3930	-12.7574
3148	18	51	20.61	-32	15	0.23	282.8359	-32.2501
3149	18	52	0.95	-16	51	55.27	283.0040	-16.8654
3150	18	53	16.41	-32	41	23.88	283.3184	-32.6900
3151	18	56	31.42	-34	20	22.57	284.1309	-34.3396
3152	18	58	21.66	-23	38	40.65	284.5902	-23.6446
3153	18	58	40.75	-36	5	48.91	284.6698	-36.0969
3154	18	59	15.68	-32	50	48.29	284.8153	-32.8467
3155	18	59	17.15	-20	3	40.77	284.8215	-20.0613
3156	19	1	29.14	-34	57	58.6	285.3714	-34.9663
3157	19	1	41.71	-24	12	47.32	285.4238	-24.2131
3158	19	3	42.88	-24	33	59.66	285.9287	-24.5666
3159	19	4	21.22	-20	46	4.77	286.0884	-20.7680
3160	19	4	23.35	-21	34	33.05	286.0973	-21.5758
3161	19	12	24.11	-26	37	16.11	288.1005	-26.6211

							RA DECIMAL	DEC DECIMAL
3162	19	13	36.41	-18	5	42.41	288.4017	-18.0951
3163	19	13	45.37	-14	17	47.63	288.4391	-14.2966
3164	19	17	33.29	-18	39	34.97	289.3887	-18.6597
3165	19	17	38.94	-14	55	48.37	289.4123	-14.9301
3166	19	17	58.02	-15	37	11.41	289.4918	-15.6198
3167	19	18	1.55	-24	37	58.11	289.5065	-24.6328
3168	19	18	4.73	-14	42	57.25	289.5197	-14.7159
3169	19	18	5.2	-15	39	28.35	289.5217	-15.6579
3170	19	19	41.15	-16	57	3.62	289.9215	-16.9510
3171	19	23	47.03	-19	39	23.79	290.9459	-19.6566
3172	19	24	29.37	-25	6	50.77	291.1224	-25.1141
3173	19	31	33.22	-19	13	16.69	292.8884	-19.2213
3174	19	38	6.41	-17	33	55.98	294.5267	-17.5656
3175	19	40	59.57	-16	20	22.57	295.2482	-16.3396
3176	20	0	35.95	-32	40	50.25	300.1498	-32.6806
3177	20	20	33.54	-29	10	15	305.1398	-29.1708

TAURUS

#	H	M	S	D	M	S	RA DECIMAL	DEC DECIMAL
3178	5	29	51.39	20	2	56.53	82.4641	20.0490
3179	5	34	45.3	27	5	26.36	83.6888	27.0907
3180	5	35	10.73	20	14	8.58	83.7947	20.2357
3181	5	41	11.87	20	48	0.28	85.2995	20.8001
3182	5	41	53.8	17	3	40.44	85.4741	17.0612
3183	5	42	19.04	25	46	57.81	85.5793	25.7827
3184	5	44	30.68	18	35	39.21	86.1278	18.5942
3185	5	46	3.87	21	17	7.3	86.5161	21.2854
3186	5	48	15.82	21	5	43.67	87.0659	21.0955
3187	5	51	50.5	16	46	7.79	87.9604	16.7688
3188	5	52	12.84	13	23	30.05	88.0535	13.3917

TELESCOPIUM

#	H	M	S	D	M	S	RA DECIMAL	DEC DECIMAL
3189	18	3	8.62	-54	31	52.65	270.7859	-54.5313
3190	18	6	4.78	-46	5	5.39	271.5199	-46.0848
3191	18	7	49.62	-47	31	54.72	271.9567	-47.5319
3192	18	14	31.55	-46	0	15.99	273.6315	-46.0044
3193	18	17	38.26	-44	33	48.5	274.4094	-44.5635
3194	18	18	58.38	-47	7	47.75	274.7432	-47.1299
3195	18	20	7.64	-48	6	50.46	275.0318	-48.1140
3196	18	20	36.24	-48	20	41.44	275.1510	-48.3448
3197	18	22	31.52	-47	23	51.16	275.6313	-47.3975
3198	18	27	30.86	-53	4	16.87	276.8786	-53.0714
3199	18	31	22.18	-48	20	57.89	277.8424	-48.3494
3200	18	35	10.86	-46	44	55.44	278.7953	-46.7487
3201	18	36	52.25	-43	13	3.94	279.2177	-43.2178
3202	19	24	40.63	-51	26	45.14	291.1693	-51.4459

TRIANGULUM AUSTRALE

#	H	M	S	D	M	S	RA DECIMAL	DEC DECIMAL
3203	14	35	36.33	-65	21	13.29	218.9014	-65.3537
3204	14	35	43.32	-64	52	19.68	218.9305	-64.8721
3205	14	37	27.74	-65	26	54.32	219.3656	-65.4484
3206	14	42	14.56	-66	16	53.29	220.5607	-66.2815
3207	14	43	21.99	-65	38	38.95	220.8416	-65.6442
3208	14	43	54.6	-64	59	2.26	220.9775	-64.9840
3209	14	51	59.98	-63	54	7.14	222.9999	-63.9020
3210	14	53	33.09	-67	42	47.31	223.3879	-67.7131
3211	15	0	49.39	-65	25	38.42	225.2058	-65.4273
3212	15	1	41.03	-67	13	22.07	225.4210	-67.2228
3213	15	2	9.13	-67	7	35.83	225.5381	-67.1266
3214	15	3	41.54	-66	10	31.26	225.9231	-66.1753
3215	15	4	5.67	-67	25	24.57	226.0236	-67.4235
3216	15	4	36.16	-65	23	34.31	226.1507	-65.3929
3217	15	5	52.59	-60	43	59.9	226.4691	-60.7333
3218	15	6	21.41	-62	11	11	226.5892	-62.1864
3219	15	6	26.83	-62	8	7.12	226.6118	-62.1353
3220	15	8	14.71	-67	0	39.18	227.0613	-67.0109
3221	15	8	43.53	-67	52	7.99	227.1814	-67.8689

3222	15	12	46.05	-60	23	30.46	228.1919	-60.3918
3223	15	16	33.53	-67	31	24.86	229.1397	-67.5236
3224	15	23	45.18	-63	56	28.09	230.9383	-63.9411
3225	15	24	21.94	-61	35	7.48	231.0914	-61.5854
3226	15	24	53.23	-64	50	2.37	231.2218	-64.8340
3227	15	26	30.44	-67	46	32.39	231.6268	-67.7757
3228	15	27	21.31	-59	32	2.45	231.8388	-59.5340
3229	15	29	15.29	-68	22	45.55	232.3137	-68.3793
3230	15	31	35.92	-63	51	6.77	232.8997	-63.8519
3231	15	33	6.78	-59	55	11.48	233.2783	-59.9199
3232	15	33	29.79	-59	52	21.26	233.3741	-59.8726
3233	15	33	34.13	-66	9	31.71	233.3922	-66.1588
3234	15	34	47.33	-65	7	21.54	233.6972	-65.1226
3235	15	35	17.9	-68	36	2.18	233.8246	-68.6006
3236	15	38	0.89	-65	30	30.34	234.5037	-65.5084
3237	15	38	3.74	-66	15	23.63	234.5156	-66.2566
3238	15	38	6.12	-65	27	5.06	234.5255	-65.4514
3239	15	38	13.99	-65	27	54.98	234.5583	-65.4653
3240	15	38	27.1	-65	27	45.61	234.6129	-65.4627
3241	15	38	30.93	-66	17	39.13	234.6289	-66.2942
3242	15	38	38.38	-65	26	10.14	234.6599	-65.4362
3243	15	39	52.12	-60	22	12.55	234.9672	-60.3702
3244	15	40	58.84	-65	21	49.98	235.2452	-65.3639
3245	15	41	3.77	-68	27	25.95	235.2657	-68.4572
3246	15	41	7.97	-65	26	15.56	235.2832	-65.4377
3247	15	41	45.13	-68	1	8.2	235.4381	-68.0189
3248	15	42	14.12	-68	3	40.19	235.5589	-68.0612
3249	15	42	17.44	-63	49	31.39	235.5727	-63.8254
3250	15	42	18.04	-61	53	24.71	235.5752	-61.8902
3251	15	42	35.97	-62	34	56.41	235.6499	-62.5823
3252	15	44	10.56	-62	7	55.95	236.0440	-62.1322
3253	15	45	18.07	-59	10	10.9	236.3253	-59.1697
3254	15	45	28.65	-62	34	11	236.3694	-62.5697
3255	15	45	28.73	-59	9	2.94	236.3697	-59.1508
3256	15	46	18.11	-65	22	18.99	236.5754	-65.3719
3257	15	47	58.84	-62	43	55.61	236.9952	-62.7321
3258	15	50	10.65	-65	13	17.58	237.5444	-65.2216
3259	15	50	35.78	-66	17	45.05	237.6491	-66.2958
3260	15	50	50.15	-59	57	8.75	237.7090	-59.9524
3261	15	53	4.65	-64	11	7.74	238.2694	-64.1855
3262	15	53	58.64	-64	8	6.62	238.4944	-64.1352
3263	15	54	50.26	-63	18	26.89	238.7094	-63.3075
3264	15	55	7.95	-63	18	23.36	238.7831	-63.3065

3265	15	56	11.53	-67	23	53.25	239.0480	-67.3981
3266	15	58	14.08	-62	8	13.47	239.5586	-62.1371
3267	15	58	45.99	-66	52	48.51	239.6916	-66.8801
3268	15	59	1.78	-64	4	16.5	239.7574	-64.0712
3269	15	59	5.49	-66	55	6.59	239.7729	-66.9185
3270	15	59	31.37	-59	2	50.04	239.8807	-59.0472
3271	15	59	43.83	-59	5	58.65	239.9326	-59.0996
3272	16	0	39.32	-67	23	25.94	240.1638	-67.3905
3273	16	1	6.28	-59	22	53.89	240.2762	-59.3816
3274	16	2	43.69	-64	31	23.38	240.6821	-64.5232
3275	16	4	23.52	-63	7	6.82	241.0980	-63.1186
3276	16	4	27.44	-61	48	12.59	241.1144	-61.8035
3277	16	4	41.07	-59	10	33.34	241.1711	-59.1759
3278	16	5	16.8	-66	28	51.17	241.3200	-66.4809
3279	16	5	18.75	-62	7	24.62	241.3281	-62.1235
3280	16	6	51.06	-63	19	0.05	241.7127	-63.3167
3281	16	7	10.63	-64	7	38.25	241.7943	-64.1273
3282	16	7	37.82	-60	4	48.96	241.9076	-60.0803
3283	16	7	58.75	-66	15	36.52	241.9948	-66.2601
3284	16	8	34.1	-68	31	53.98	242.1421	-68.5317
3285	16	8	42.68	-62	20	25.68	242.1778	-62.3405
3286	16	8	45.31	-63	33	10.69	242.1888	-63.5530
3287	16	8	47.19	-60	23	5.36	242.1966	-60.3848
3288	16	8	51.17	-63	31	21.57	242.2132	-63.5227
3289	16	10	16.71	-62	45	43.85	242.5696	-62.7622
3290	16	10	23.43	-59	59	14.34	242.5976	-59.9873
3291	16	11	22.85	-64	2	14.89	242.8452	-64.0375
3292	16	11	26.73	-66	58	42.06	242.8614	-66.9783
3293	16	11	31.05	-61	49	2.86	242.8794	-61.8175
3294	16	11	31.96	-66	58	19.31	242.8831	-66.9720
3295	16	11	38.47	-61	4	15.42	242.9103	-61.0710
3296	16	14	2.14	-70	14	16.56	243.5089	-70.2379
3297	16	14	22.18	-66	5	3.32	243.5924	-66.0843
3298	16	14	42.1	-66	7	48.55	243.6754	-66.1302
3299	16	14	46.75	-63	47	10.79	243.6948	-63.7863
3300	16	15	47.08	-60	14	15.45	243.9461	-60.2376
3301	16	16	11.26	-62	10	11.18	244.0469	-62.1698
3302	16	16	30.56	-65	47	47.8	244.1274	-65.7966
3303	16	17	43.13	-66	19	54.89	244.4297	-66.3319
3304	16	19	4.61	-61	58	3.3	244.7692	-61.9676
3305	16	20	1.23	-62	54	32.19	245.0051	-62.9089
3306	16	20	21.12	-61	0	47.1	245.0880	-61.0131
3307	16	20	28.13	-61	3	38.35	245.1172	-61.0607

3308	16	20	34.07	-68	32	11.93	245.1420	-68.5366
3309	16	20	44.58	-65	50	45.33	245.1857	-65.8459
3310	16	22	34.91	-62	55	22.84	245.6454	-62.9230
3311	16	23	33.31	-60	4	17.26	245.8888	-60.0715
3312	16	23	35.72	-62	42	8.01	245.8989	-62.7022
3313	16	23	37.9	-64	24	27.26	245.9079	-64.4076
3314	16	25	1.83	-59	35	54.25	246.2576	-59.5984
3315	16	25	57.59	-59	25	26.41	246.4899	-59.4240
3316	16	26	33.93	-63	23	25.05	246.6414	-63.3903
3317	16	28	43.92	-61	26	12.86	247.1830	-61.4369
3318	16	28	59.28	-62	41	3.33	247.2470	-62.6843
3319	16	29	2.98	-60	30	42.7	247.2624	-60.5119
3320	16	29	4.44	-62	36	57.7	247.2685	-62.6160
3321	16	29	21.38	-63	50	40.74	247.3391	-63.8446
3322	16	31	9.7	-58	48	42.14	247.7904	-58.8117
3323	16	31	21.45	-68	15	20.54	247.8394	-68.2557
3324	16	31	35.68	-58	48	7.66	247.8987	-58.8021
3325	16	33	42.07	-63	44	35.35	248.4253	-63.7432
3326	16	34	8.3	-63	43	18.64	248.5346	-63.7218
3327	16	35	25.58	-64	36	31.38	248.8566	-64.6087
3328	16	35	36.23	-64	37	35.56	248.9009	-64.6265
3329	16	35	54.31	-64	34	10.13	248.9763	-64.5695
3330	16	37	40.79	-61	35	7.7	249.4200	-61.5855
3331	16	38	10.43	-66	23	13.68	249.5434	-66.3871
3332	16	39	3.08	-62	16	5.09	249.7628	-62.2681
3333	16	39	11.83	-68	37	28.63	249.7993	-68.6246
3334	16	40	32.95	-63	19	17.31	250.1373	-63.3215
3335	16	40	52.08	-61	8	43.09	250.2170	-61.1453
3336	16	44	34	-63	14	10.5	251.1417	-63.2363
3337	16	48	46.39	-68	4	55.92	252.1933	-68.0822

VELA

#	H	M	S	D	M	S	RA DECIMAL	DEC DECIMAL
3338	7	53	20.64	-48	54	27.91	118.3360	-48.9078
3339	8	7	8.55	-45	54	58.7	121.7856	-45.9163
3340	8	7	13.69	-48	47	45.62	121.8070	-48.7960
3341	8	7	29.75	-47	58	13.55	121.8740	-47.9704
3342	8	8	20.7	-46	11	9.13	122.0862	-46.1859
3343	8	10	4.08	-44	25	7.73	122.5170	-44.4188
3344	8	11	17.33	-45	49	23.57	122.8222	-45.8232
3345	8	12	19.07	-46	12	19.58	123.0794	-46.2054
3346	8	15	6.8	-46	8	54.72	123.7783	-46.1485
3347	8	16	21.27	-47	14	10.27	124.0886	-47.2362
3348	8	18	4.13	-46	48	25.23	124.5172	-46.8070
3349	8	18	47.16	-46	3	34.27	124.6965	-46.0595
3350	8	19	31.96	-48	12	10.28	124.8832	-48.2029
3351	8	21	22.04	-47	2	26.93	125.3418	-47.0408
3352	8	22	40.35	-44	47	29.43	125.6681	-44.7915
3353	8	22	44.01	-48	9	3.59	125.6834	-48.1510
3354	8	22	56.4	-48	13	19.19	125.7350	-48.2220
3355	8	24	42.31	-48	57	7.75	126.1763	-48.9522
3356	8	25	45.7	-45	36	59.13	126.4404	-45.6164
3357	8	28	15.28	-47	48	43.34	127.0637	-47.8120
3358	8	28	37.99	-46	55	7.34	127.1583	-46.9187
3359	8	29	33.5	-47	56	46.15	127.3896	-47.9462
3360	8	29	34.07	-48	36	20.59	127.3920	-48.6057
3361	8	29	46.43	-47	49	25.75	127.4434	-47.8238
3362	8	31	3.45	-40	17	36.1	127.7644	-40.2934
3363	8	32	17.98	-39	48	27.95	128.0749	-39.8078
3364	8	33	11.05	-43	28	0.5	128.2960	-43.4668
3365	8	34	27.65	-45	37	17.16	128.6152	-45.6214
3366	8	34	48.41	-42	45	58.3	128.7017	-42.7662
3367	8	37	28.68	-41	34	51.98	129.3695	-41.5811
3368	8	46	37.1	-38	52	12.34	131.6546	-38.8701
3369	8	48	39.2	-39	6	4.94	132.1633	-39.1014
3370	8	48	49.54	-49	41	57.89	132.2064	-49.6994
3371	8	49	55.17	-47	38	3.06	132.4799	-47.6342
3372	8	50	15.4	-48	40	56.03	132.5642	-48.6822
3373	8	50	16.09	-45	23	2.46	132.5670	-45.3840
3374	8	50	27.71	-54	25	16.74	132.6154	-54.4213
3375	8	50	49.12	-46	6	28.2	132.7047	-46.1078
3376	8	51	4.55	-46	19	32.44	132.7690	-46.3257

3377	8	52	0.65	-40	22	59.53	133.0027	-40.3832
3378	8	52	1.52	-50	51	33.56	133.0063	-50.8593
3379	8	53	44.2	-54	1	53.05	133.4342	-54.0314
3380	8	54	36.82	-53	21	13.48	133.6534	-53.3537
3381	8	57	6.38	-49	45	57.64	134.2766	-49.7660
3382	8	59	50.94	-43	20	43.17	134.9622	-43.3453
3383	9	0	27.15	-54	6	41.14	135.1131	-54.1114
3384	9	2	33.29	-46	29	14.65	135.6387	-46.4874
3385	9	4	10.75	-53	28	10.68	136.0448	-53.4696
3386	9	4	56.36	-53	37	34.64	136.2348	-53.6263
3387	9	5	49.1	-42	10	8.26	136.4546	-42.1690
3388	9	6	45.91	-53	19	40.65	136.6913	-53.3280
3389	9	8	17.82	-43	21	17.39	137.0743	-43.3548
3390	9	10	15.28	-53	4	31.12	137.5636	-53.0753
3391	9	11	3.88	-53	0	0.66	137.7662	-53.0002
3392	9	13	59.74	-55	35	20.93	138.4989	-55.5891
3393	9	14	10.89	-45	3	58.62	138.5454	-45.0663
3394	9	14	34.8	-53	28	28.82	138.6450	-53.4747
3395	9	15	53.14	-52	45	52.28	138.9714	-52.7645
3396	9	16	43.5	-56	41	14.28	139.1813	-56.6873
3397	9	16	57.63	-53	52	48.75	139.2401	-53.8802
3398	9	17	30.3	-38	36	43.61	139.3763	-38.6121
3399	9	18	2.13	-39	23	8.21	139.5089	-39.3856
3400	9	18	7.79	-56	16	4.9	139.5325	-56.2680
3401	9	19	10.96	-39	40	29.81	139.7957	-39.6749
3402	9	19	21.53	-55	17	47.14	139.8397	-55.2964
3403	9	20	29.5	-53	56	12.64	140.1229	-53.9368
3404	9	23	1.21	-53	59	31.68	140.7550	-53.9921
3405	9	23	33.49	-55	36	11.64	140.8895	-55.6032
3406	9	24	15.57	-41	4	39.93	141.0649	-41.0778
3407	9	24	49.11	-37	2	53.4	141.2046	-37.0482
3408	9	25	58.12	-42	1	17.29	141.4922	-42.0215
3409	9	27	1.03	-44	15	50.96	141.7543	-44.2642
3410	9	31	13.12	-55	34	0.1	142.8047	-55.5667
3411	9	31	26.38	-55	28	22.07	142.8599	-55.4728
3412	9	32	24.31	-57	7	55.86	143.1013	-57.1322
3413	9	37	6.35	-57	4	0.56	144.2765	-57.0668
3414	9	37	18.6	-51	14	50.33	144.3275	-51.2473
3415	9	37	34.93	-44	16	0.06	144.3955	-44.2667
3416	9	37	54.86	-55	35	27.72	144.4786	-55.5910
3417	9	38	10.39	-54	56	51.07	144.5433	-54.9475
3418	9	38	44.16	-57	35	26.05	144.6840	-57.5906
3419	9	40	10.1	-43	0	46.26	145.0421	-43.0129

3420	9	40	42.65	-51	19	33.38	145.1777	-51.3259
3421	9	42	36.8	-57	44	11.92	145.6533	-57.7366
3422	9	45	48.38	-52	28	8.9	146.4516	-52.4691
3423	9	46	31.4	-45	1	19.83	146.6308	-45.0222
3424	9	46	59.63	-56	57	58.28	146.7484	-56.9662
3425	9	47	12.72	-55	50	33.11	146.8030	-55.8425
3426	9	48	31.73	-58	1	51.69	147.1322	-58.0310
3427	9	48	39.04	-58	0	35.29	147.1627	-58.0098
3428	9	48	48.32	-58	0	59.29	147.2013	-58.0165
3429	9	49	8.65	-51	9	43.06	147.2860	-51.1620
3430	9	50	31.72	-57	20	29.44	147.6322	-57.3415
3431	9	51	56.59	-45	13	17.33	147.9858	-45.2215
3432	9	52	21.04	-57	49	18.64	148.0877	-57.8218
3433	9	52	38.75	-48	48	16.81	148.1615	-48.8047
3434	9	53	13.57	-54	13	11.23	148.3065	-54.2198
3435	9	53	21.18	-52	1	27.05	148.3382	-52.0242
3436	9	53	24.73	-55	9	39.62	148.3530	-55.1610
3437	9	53	26.98	-48	13	37.21	148.3624	-48.2270
3438	9	53	36.77	-52	45	56.54	148.4032	-52.7657
3439	9	53	38.33	-55	38	51.98	148.4097	-55.6478
3440	9	53	39.55	-55	8	31.47	148.4148	-55.1421
3441	9	53	46.56	-55	5	1.89	148.4440	-55.0839
3442	9	54	36.93	-53	56	32.26	148.6539	-53.9423
3443	9	54	37.57	-55	8	27.54	148.6566	-55.1410
3444	9	54	42.26	-54	11	7.29	148.6761	-54.1854
3445	9	54	47.02	-55	8	48.95	148.6959	-55.1469
3446	9	55	7.62	-43	36	3.53	148.7817	-43.6010
3447	9	56	17.47	-54	3	23.94	149.0728	-54.0566
3448	9	56	29.73	-54	7	31.28	149.1239	-54.1254
3449	9	56	30.33	-55	57	21.03	149.1264	-55.9558
3450	9	56	31.75	-54	4	52.5	149.1323	-54.0813
3451	9	57	5.1	-56	20	6.52	149.2713	-56.3351
3452	9	57	13.84	-55	32	5.57	149.3077	-55.5349
3453	9	57	30.94	-56	3	46.5	149.3789	-56.0629
3454	10	0	5.75	-54	0	16.54	150.0240	-54.0046
3455	10	0	15.78	-54	0	20.94	150.0657	-54.0058
3456	10	1	47.63	-43	34	51.12	150.4485	-43.5809
3457	10	3	55.35	-56	43	2.82	150.9806	-56.7174
3458	10	4	3.06	-46	10	48.06	151.0128	-46.1800
3459	10	4	8.15	-44	57	7.98	151.0340	-44.9522
3460	10	4	33.25	-51	2	40.29	151.1385	-51.0445
3461	10	5	30.84	-52	34	16.25	151.3785	-52.5712
3462	10	5	51.23	-44	52	48.6	151.4635	-44.8802

3463	10	6	38.01	-46	46	23.42	151.6584	-46.7732
3464	10	7	55.61	-45	5	17.8	151.9817	-45.0883
3465	10	10	3.19	-54	43	21.05	152.5133	-54.7225
3466	10	10	35.34	-55	17	7.88	152.6473	-55.2855
3467	10	10	42.77	-47	34	19.47	152.6782	-47.5721
3468	10	11	12.08	-54	54	31.15	152.8004	-54.9087
3469	10	15	8.71	-49	18	10.85	153.7863	-49.3030
3470	10	15	51.72	-49	4	50.42	153.9655	-49.0807
3471	10	16	7.03	-53	57	46.01	154.0293	-53.9628
3472	10	16	42.12	-52	25	59.51	154.1755	-52.4332
3473	10	17	10.07	-51	31	34.93	154.2919	-51.5264
3474	10	18	38.98	-54	50	56.34	154.6624	-54.8490
3475	10	18	52.08	-54	55	12	154.7170	-54.9200
3476	10	19	10.49	-51	16	6.04	154.7937	-51.2683
3477	10	19	22.12	-51	16	26.43	154.8422	-51.2740
3478	10	21	40.01	-50	28	2.8	155.4167	-50.4674
3479	10	21	54.92	-51	6	12.88	155.4788	-51.1036
3480	10	23	3.25	-49	44	10.26	155.7635	-49.7362
3481	10	23	11.16	-53	28	29.08	155.7965	-53.4747
3482	10	23	38.54	-43	56	49.61	155.9106	-43.9471
3483	10	24	34.14	-54	0	53.57	156.1423	-54.0149
3484	10	26	23.06	-54	34	8.23	156.5961	-54.5690
3485	10	27	24.77	-47	28	43.49	156.8532	-47.4787
3486	10	29	3.02	-54	43	55.23	157.2626	-54.7320
3487	10	29	7.09	-48	56	20.59	157.2795	-48.9391
3488	10	29	12.82	-47	50	27.86	157.3034	-47.8411
3489	10	29	21.67	-54	41	48.28	157.3403	-54.6967
3490	10	33	12.43	-44	31	36	158.3018	-44.5267
3491	10	33	56.45	-51	9	2.78	158.4852	-51.1508
3492	10	33	56.85	-55	47	29.59	158.4869	-55.7916
3493	10	33	58.35	-53	13	30.33	158.4931	-53.2251
3494	10	34	2.32	-51	5	56.68	158.5097	-51.0991
3495	10	34	9.62	-51	3	11.44	158.5401	-51.0532
3496	10	34	48.64	-54	57	30.48	158.7026	-54.9585
3497	10	35	20.64	-53	52	12.95	158.8360	-53.8703
3498	10	35	30.94	-51	59	9.74	158.8789	-51.9860
3499	10	35	42.44	-53	12	13.52	158.9269	-53.2038
3500	10	36	55.48	-55	19	56.13	159.2312	-55.3323
3501	10	37	5.07	-50	43	36.33	159.2711	-50.7268
3502	10	37	37.91	-54	55	39.18	159.4080	-54.9275
3503	10	38	6.95	-55	32	25.98	159.5289	-55.5405
3504	10	38	7.92	-46	15	44.29	159.5330	-46.2623
3505	10	38	32.34	-56	1	52.55	159.6347	-56.0313

3506	10	39	19.38	-54	47	59.4	159.8308	-54.7998
3507	10	40	36.45	-46	8	33.18	160.1519	-46.1426
3508	10	40	41.36	-46	8	12.69	160.1723	-46.1369
3509	10	41	14	-42	23	56.5	160.3083	-42.3990
3510	10	42	33	-45	12	28.79	160.6375	-45.2080
3511	10	42	39.36	-54	35	4.12	160.6640	-54.5845
3512	10	43	10.17	-47	8	16.58	160.7924	-47.1379
3513	10	43	22.08	-47	10	13.79	160.8420	-47.1705
3514	10	44	9.54	-53	3	22.22	161.0398	-53.0562
3515	10	44	18.63	-53	6	13.48	161.0776	-53.1037
3516	10	47	36.76	-54	57	16.71	161.9032	-54.9546
3517	10	48	20.69	-52	20	50.76	162.0862	-52.3474
3518	10	48	26.99	-55	13	16.25	162.1125	-55.2212
3519	10	49	41.41	-53	52	50.73	162.4226	-53.8808
3520	10	49	43.24	-50	14	21.55	162.4302	-50.2393
3521	10	49	44.97	-54	37	31.97	162.4374	-54.6255
3522	10	50	19.82	-55	12	29.99	162.5826	-55.2083
3523	10	51	38.19	-51	46	2.63	162.9091	-51.7674
3524	10	52	6.9	-52	57	43.7	163.0288	-52.9621
3525	10	52	20.86	-52	58	14.94	163.0869	-52.9708
3526	10	52	27.97	-53	2	45.92	163.1165	-53.0461
3527	10	53	37.25	-55	24	5.29	163.4052	-55.4015
3528	10	54	48.45	-51	6	12.92	163.7019	-51.1036
3529	10	55	16.71	-55	10	18.28	163.8196	-55.1717
3530	10	56	35	-55	13	11.43	164.1458	-55.2198
3531	11	0	44.83	-54	51	41.99	165.1868	-54.8617
3532	11	0	54.69	-54	51	1.44	165.2279	-54.8504
3533	11	1	38.91	-53	35	10.44	165.4121	-53.5862
3534	11	2	29.97	-48	49	0.91	165.6249	-48.8169
3535	11	3	4.91	-54	19	49.37	165.7705	-54.3304
3536	11	6	31.05	-53	16	6.25	166.6294	-53.2684
3537	11	8	37.36	-52	59	29.42	167.1557	-52.9915

VOLANS

#	H	M	S	D	M	S	RA DECIMAL	DEC DECIMAL
3538	6	42	16.58	-64	31	19.42	100.5691	-64.5221
3539	6	56	47.41	-64	34	26.1	104.1975	-64.5739

Nicholson #3539 – 15 x 15 arc minutes

VULPECULA

#	H	M	S	D	M	S	RA DECIMAL	DEC DECIMAL
3540	18	58	32.5	21	12	14.24	284.6354	21.2040
3541	19	0	33.19	20	14	25.9	285.1383	20.2405
3542	19	4	4.59	18	20	7.46	286.0191	18.3354
3543	19	7	51.49	21	15	43.51	286.9646	21.2621
3544	19	16	8.3	24	46	44.32	289.0346	24.7790
3545	19	29	5.02	24	38	9.64	292.2709	24.6360
3546	19	30	30.39	28	21	51.19	292.6266	28.3642
3547	19	30	36.79	28	20	44.63	292.6533	28.3457
3548	19	32	31.3	25	32	10.52	293.1304	25.5363
3549	19	32	32.35	29	51	4.83	293.1348	29.8513
3550	19	33	56.23	28	17	14.78	293.4843	28.2874
3551	19	34	32.53	29	39	12.43	293.6356	29.6535
3552	19	36	16.25	29	26	59.72	294.0677	29.4499
3553	19	36	21.42	29	24	56.39	294.0893	29.4157
3554	19	36	23.05	29	21	26.56	294.0960	29.3574

3555	19	36	27.13	29	25	44.66	294.1131	29.4291
3556	19	36	32.99	26	43	31.2	294.1375	26.7253
3557	19	36	39.77	29	18	51.53	294.1657	29.3143
3558	19	36	48.73	30	42	21.81	294.2031	30.7061
3559	19	36	59.19	29	43	5.82	294.2466	29.7183
3560	19	37	2.23	28	8	57.97	294.2593	28.1494
3561	19	37	14.06	28	44	1.07	294.3086	28.7336
3562	19	37	34.95	28	17	14.17	294.3956	28.2873
3563	19	37	43.34	29	36	25.43	294.4306	29.6071
3564	19	37	45.75	28	19	42.41	294.4406	28.3284
3565	19	38	11.87	26	44	31.36	294.5495	26.7420
3566	19	38	26.82	30	27	43.61	294.6118	30.4621
3567	19	38	29.06	30	28	5.63	294.6211	30.4682
3568	19	38	33.92	30	25	33.32	294.6413	30.4259
3569	19	39	46.22	29	41	30.26	294.9426	29.6917
3570	19	43	12.33	17	49	17.44	295.8014	17.8215
3571	19	43	43.06	18	5	36.58	295.9294	18.0935
3572	19	45	3.64	18	59	39.45	296.2651	18.9943
3573	19	45	17.67	18	48	30.61	296.3236	18.8085
3574	19	45	26.17	28	45	7.25	296.3590	28.7520
3575	19	50	28.08	19	58	25.04	297.6170	19.9736
3576	19	50	41.96	20	18	29.21	297.6748	20.3081
3577	19	50	48.8	20	15	21.17	297.7033	20.2559
3578	19	52	27.56	27	28	33.77	298.1148	27.4760
3579	19	52	37.48	26	37	33.29	298.1562	26.6259
3580	19	53	9.79	28	36	30.09	298.2908	28.6084
3581	19	53	32.29	27	12	37.83	298.3845	27.2105
3582	19	53	46.99	23	27	52.86	298.4458	23.4647
3583	19	53	54.97	20	29	48.28	298.4790	20.4967
3584	19	54	40.07	21	8	0.17	298.6670	21.1334
3585	19	55	47.43	28	27	33.21	298.9476	28.4592
3586	19	59	54.4	27	32	13.18	299.9767	27.5370
3587	20	0	12.66	29	45	39.95	300.0527	29.7611
3588	20	0	25.14	29	40	1.18	300.1047	29.6670
3589	20	0	28.81	29	43	16.9	300.1201	29.7214
3590	20	0	51.22	29	50	44.8	300.2134	29.8458
3591	20	1	27.9	30	1	49.88	300.3662	30.0305
3592	20	7	23.47	26	33	5.97	301.8478	26.5517
3593	20	10	33.07	29	32	3.04	302.6378	29.5342
3594	20	14	54.52	25	18	11.65	303.7272	25.3032
3595	20	25	23.54	27	30	51.11	306.3481	27.5142
3596	20	25	24.1	23	54	51.09	306.3504	23.9142
3597	20	28	57.67	24	6	53.98	307.2403	24.1150

3598	20	29	19.5	26	33	51.61	307.3313	26.5643
3599	20	29	56.13	23	10	45.68	307.4839	23.1794
3600	20	31	8.58	26	41	56.05	307.7858	26.6989
3601	20	31	56.26	23	56	13.06	307.9844	23.9370
3602	20	33	48.08	21	9	37.7	308.4503	21.1605
3603	20	36	7.28	25	36	48.39	309.0303	25.6134
3604	20	37	7.35	29	53	37.26	309.2806	29.8937
3605	20	37	27.42	26	45	2.26	309.3642	26.7506
3606	20	38	18.71	27	8	4.47	309.5780	27.1346
3607	20	39	51.04	20	23	22.75	309.9627	20.3897
3608	20	40	41.2	21	3	30	310.1717	21.0583
3609	20	40	55.26	22	33	29.41	310.2303	22.5582
3610	20	41	52.12	28	15	35.32	310.4672	28.2598
3611	20	47	39.52	27	7	33.51	311.9147	27.1260
3612	20	49	9.51	25	41	56.59	312.2896	25.6991
3613	20	50	55.78	26	33	22.43	312.7324	26.5562
3614	21	17	7.58	28	14	26.18	319.2816	28.2406
3615	21	17	16.63	28	12	9.75	319.3193	28.2027
3616	21	23	36.41	29	51	16.75	320.9017	29.8547

Nicholson #3616 – 15 x 15 arc minutes

BY THE SAME AUTHOR

All are available from Amazon.com and from Amazon.co.uk

1800 new double stars for amateur observers

Discover your own double star

Discover your own variable star

Identifying Common Proper Motion Binary Star Systems

Identifying Identical Twin Star Systems from the SDSS Data Release 10

www.ingramcontent.com/pod-product-compliance
Lightning Source LLC
Chambersburg PA
CBHW080304180526
45167CB00006B/2660